Corrosion and Corrosion Control

Corrosion and Corrosion Control

Editor

Vikas Saini

scitus
academics

Corrosion and Corrosion Control

Edited by **Vikas Saini**

Printed in 2017

ISBN: 978-1-68117-213-2

Library of Congress Control Number: 2015936575

Contents

Preface

Corrosion is the gradual destruction of materials (usually metals) by chemical reaction with their environment. In the most common use of the word, this means electrochemical oxidation of metals in reaction with an oxidant such as oxygen. Rusting, the formation of iron oxides, is a well-known example of electrochemical corrosion. This type of damage typically produces oxide(s) or salt(s) of the original metal. Corrosion can also occur in materials other than metals, such as ceramics or polymers, although in this context, the term degradation is more common. Corrosion degrades the useful properties of materials and structures including strength, appearance and permeability to liquids and gases.

Editor

Corrosion Control of Aluminum Surfaces by Polypyrrole Films: Influence of Electrolyte

Andréa Santos Liu[I, II] and Maria Auxiliadora Silva Oliveira[I]

[I]Departamento de Química, Instituto Tecnológico de Aeronáutica – ITA, Centro Técnico Aeroespacial – CTA, Praça Marechal Eduardo Gomes, 50, 12228-900 São José dos Campos - SP, Brazil

[II]Instituto de Pesquisa e Desenvolvimento, Universidade do Vale do Paraíba – UNIVAP, Av. Shishima Hifumi, 2911, 12244-000 São José dos Campos - SP, Brazil

ABSTRACT

Polypyrrole (PPy) films were galvanostatically deposited on 99.9 wt. (%) aluminum electrodes from aqueous solutions containing each carboxylic acid: tartaric, oxalic or citric. Scanning Electron Microscopy (SEM) was used to analyze the morphology of the aluminum surfaces coated with the polymeric films. It was observed that the films deposited from tartaric acid medium presented higher homogeneity than those deposited from oxalic and citric acid. Furthermore, the corrosion protection of aluminum surfaces by PPy films was also investigated by potentiodynamic polarization experiments.

INTRODUCTION

Polypyrrole (PPy) is a conducting polymer that has been investigated for applications in batteries, sensors, membranes and protection of metals against corrosion[1-4].

The polymer can be prepared by chemical and electrochemical methods from aqueous or organic media. The electrochemical process is more advantageous since film properties such as thickness and conductivity can be controlled by the synthesis parameters (current density, substrate, pH, nature and concentration of electrolyte)[5, 6].

The electrolytic species can participate as dopant and incorporate into the polymeric films[7]. The anions are indispensable to compensate the conducting polymer charges. The structure and the concentration of these anions affect the conductivity, the stability and the morphology of the PPy films[8]. It was found that the films doped with aromatic organic anions are electrically more conductive than those doped with inorganic anions. This behavior has been ascribed to a better superposition of the molecular orbitals of the dopant with the p atomic orbitals of the carbon in the polymeric chain. It was also observed that PPy films doped with organic species are more adherent and uniform than those doped with inorganic anions[9].

In a previous work, we have reported that adherent and homogeneous PPy films were galvanostatically electrodeposited on aluminum surfaces from aqueous solutions containing pyrrole and tartaric acid, pH 2, applying current density of 2.5 mA.cm^{-2}[10].

In this work, the influence of aliphatic organic acids on the electrodeposition of PPy films on 99.9 wt. (%) aluminum surfaces was investigated. The polymeric films were galvanostatically deposited at 2.5 mA.cm^{-2} from aqueous solutions containing, respectively tartaric, oxalic or citric acid The morphology of the PPy films was characterized by Scanning Electron Microscopy (SEM). Additionally, the efficiency of the polymeric films on protecting aluminum surfaces against corrosion was investigated by polarization curves in chloride medium.

EXPERIMENTAL

The electrochemical experiments were performed at room temperature in a cell containing three electrodes. The working electrode was a 99.9 wt. (%) aluminum rod, embedded in Teflon®, leaving a disc-shaped exposed area of 0.53 cm^2. The reference electrode was a saturated Ag / AgCl, Cl$^-$ electrode, and the auxiliary electrode was a platinum wire. These experiments were carried out by a Potentiostat / Galvanostat MQPG-01 Model (Microquímica).

The PPy films were galvanostatically deposited from aqueous solutions containing 0.5 mol.L^{-1} pyrrole (Aldrich, distilled before using) + 0.2 mol.L^{-1} of each aliphatic acid: tartaric acid (Reagen), oxalic acid (Fisher) or citric acid (Reagen). The pH of each solution was adjusted to 2.0 by addition of NaOH (Synth). The applied current density was 2.5 mA.cm^{-2}.

Before each electrochemical experiment, the aluminum surfaces were polished with emery paper (400, 600 and 1000 grades), 3-mm alumina water suspension and rinsed with distilled water.

The morphology of aluminum surfaces coated with PPy films was analyzed using a Jeol JXA-840A Scanning Electron Microscope (SEM). The micrographs were obtained using an electron beam of 15 keV.

FTIR spectroscopy was used to analyze the PPy films composition. The spectra were obtained using a spectrometer model SPECTRUM-2000 (Perkin Elmer). The analysis conditions were: wavenumber range of 4000-400 cm^{-1}, 4 cm^{-1} resolution, 40 scans, and room temperature (25 °C). KBr pellets were prepared with the PPy films. The films were dried at vacuum and then carefully removed from aluminum surfaces with aid of a spatula.

The corrosion resistance of aluminum surfaces, polished and coated with PPy films, was investigated using the potentiodynamic polarization technique. The surfaces were exposed to a 0.1 mol.L^{-1} NaCl aqueous solution (pH 5.9), not stirred and open to the atmosphere. The polarization curves were obtained starting from the open circuit potential (OCP) and varying the potential, respectively, up to 400 mV (anodic branch of the Tafel plot) and down to −400 mV (cathodic branch of the Tafel plot). The potential scan rate was 5 mV.s^{-1}.

The corrosion potential (E_{corr}) and the corrosion current densitiy (j_{corr}) were obtained from the Tafel plots. The E_{corr} is the potential at which the current density is zero. The j_{corr} was determined by extrapolation, to E_{corr}, from linear parts of the anodic and cathodic branches of the Tafel plots[11].

RESULTS AND DISCUSSION

Figure 1 presents the cronopotentiometric curves for the electrodeposition of PPy on aluminum surfaces from aqueous solutions containing 0.5 mol.L^{-1} pyrrole + 0.2 mol.L^{-1} each organic acid and pH 2.0. The applied current density was 2.5 mA.cm^{-2} and maintaining the same deposition charge (9.10^4 C.m^{-2}).

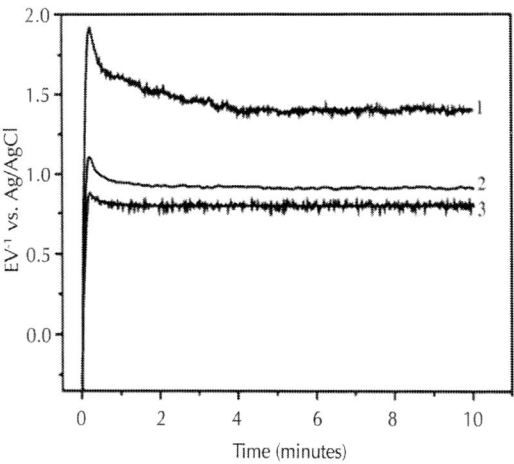

Figure 1: Potential transient for PPy electrodeposition on aluminum surfaces from aqueous solutions containing 0.5 mol.L^{-1} pyrrole and 0.2 mol.L^{-1} of each electrolyte: 1) citric acid; 2) tartaric acid; and 3) oxalic acid.

It was observed that the steady-potential value at which occurs the growth of PPy films, decreases in the following sequence: citric acid > tartaric acid > oxalic acid. This behavior has been associated to aluminum oxide layer characteristics.

The proposed mechanism to explain the PPy formation on aluminum surfaces considers the growth of polymer nucleus through pores and/ or cracks on the oxide layer[7, 12, and 13]. Consequently, a more compact oxide layer, with fewer defects, should difficult the PPy growth. The adsorption of stable complexes at the interface aluminum oxide/ electrolytic solution may result in a more compact and resistant oxide layer[14, 15]. The formation of complexes between aluminum species and the organic acids occurs through carboxylic groups[16].

Among the used aliphatic acid, the citric acid presents higher number of active sites to complexing with aluminum species at the interface oxide/solution[16]. Consequently the oxide layer is more compact and thicker in citric acid medium and can explain the PPy growth at higher potential. Figure 2 presents the structures of the organic acids used as electrolytes. Studies were being performed in our laboratory to verify the behavior of these acids in the anodization of aluminum surfaces and their role as corrosion inhibitors.

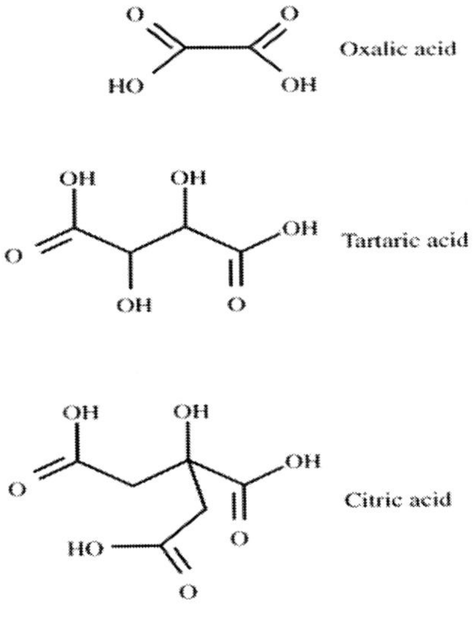

Figure 2: Structures of organic acids used as electrolytes to deposit the PPy films.

The PPy films deposited from each aliphatic acid were removed from aluminum surfaces and characterized by FTIR spectroscopy. The results are presented in Figure 3.

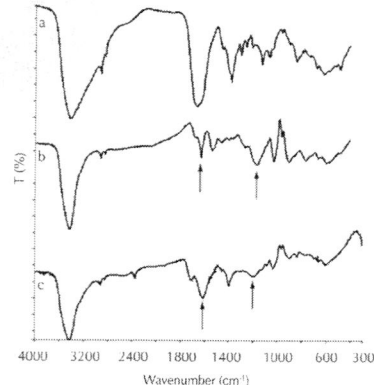

Figure 3: FTIR spectra of PPy films electrodeposited from aqueous solutions containing 0.5 mol.L^{-1} pyrrole + 0.2 mol L^{-1} of each electrolyte: a) tartaric acid; b) oxalic acid; and c) citric acid.

The absorption bands at 1170 and 1630 cm^{-1} in the FTIR spectra of PPy films electrodeposited from oxalic and citric acid media have been attributed, respectively, to the bipolaronic species and to carbonyl groups that are formed by overoxidation[17, 18]. Similar bands were not observed in the FTIR spectrum of PPy film electrodeposited from tartaric acid medium, which indicates that the polymer deposited from this medium presents lower overoxidation degree than the films formed in oxalic or citric acids.

The overoxidation is an irreversible degradation process that results in the shortening of the polymer chain length and/or formation of defects and pores along the PPy chain[19]. Figure 4 shows the sequence of reactions that occur during the overoxidation.

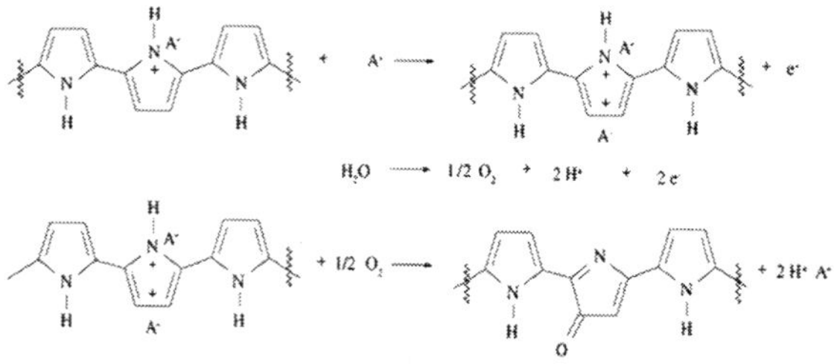

Figure 4: Reactions that occur during the overoxidation of the PPy films.

The spectrum of PPy film deposited from tartaric acid (Figure 3a) presents a very strong band at 1700 cm[-1] and a medium band at 1134 cm[-1], attributed to the carboxylic and hydroxyl groups, respectively[20, 21]. These bands are associated to the tartaric acid, which incorporates the polymeric film via doping during the synthesis of the PPy. Since the overoxidation also results in the dedoping of conducting polypyrrole[19], it can conclude that the PPy film deposited in tartaric acid is not overoxidized.

Scanning Electron Microscopy was used to investigate the morphology of the PPy films electrodeposited from each aliphatic acid. The obtained micrographs are presented in Figure 5.

Figure 5: SEM micrographs of aluminum surfaces coated with PPy films deposited galvanostatically at 2.5 mA.cm[-2] from aqueous media containing 0.5 mol.L[-1] pyrrole + 0.2 mol.L[-1] of each electrolyte: a) tartaric acid; b) oxalic acid; and c) citric acid.

The PPy films present a *cauliflower*-like structure constituted by micro-spherical grains. It has been reported that this *cauliflower* structure is related to the dopant intercalation difficulty in the disordered polymeric chain[22]. The PPy film formed in tartaric acid medium was more homogeneous than the films deposited in media containing the other acids.

Figure 5 also shows that the *cauliflower* structure in the PPy films deposited from oxalic acid was smaller than those observed in the PPy films formed in citric acid, which presented a smaller structure than the PPy electrodeposited from tartaric acid. The smallest size of *cauliflower* structure has been attributed to the occurrence of overoxidation process during the synthesis of PPy.

The corrosion resistance of aluminum surfaces coated with PPy films was investigated in chloride containing medium

Figure 6 shows the potentiodynamic polarization curves, in 0.1 mol.L^{-1} NaCl aqueous solutions, not stirred pH 5.9, for just polished aluminum surfaces and for surfaces coated with PPy films deposited galvanostatically at 2.5 mA.cm^{-2} in the different electrolytes The potential scan rate was 5 mV.s^{-1}.

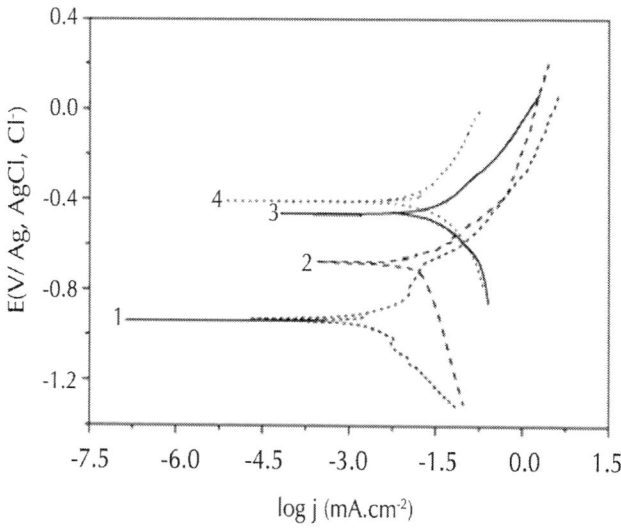

Figure 6: Polarization curves for aluminum surface: 1) just polished; coated with PPy films electrodeposited galvanostatically (2.5 mA.cm^{-2}) from pyr-role

aqueous solutions containing, respectively, 0.2 mol.L^{-1} acid: 2) citric; 3) oxalic; and 4) tartaric.

The corrosion potential of aluminum surfaces coated with polymeric films shifted to less negative values and the anodic current densities of these surfaces were smaller than those observed to just polish aluminum surfaces. These observations can be taken as an indication that aluminum surfaces coated with PPy films are more resistant to corrosion than uncoated surfaces. Furthermore, after polarization curves, it was noted pits on the uncoated aluminum surface, which presented pit potential at −0.69 V (Figure 6, curve 1). Pit potential was not observed in polarization curves for aluminum surfaces coated by PPy films.

The cathodic current densities, however, were higher for aluminum surfaces coated with PPy than for uncoated aluminum surfaces. The cathodic current densities increased in the sequence: uncoated surfaces < surface coated with PPy deposited in citric acid < surface coated with PPy deposited in tartaric acid @ surface coated with PPy deposited in oxalic acid. Similar results have been shown in literature and have been associated to reduction reaction of polymeric matrix, which contributes to increasing the cathodic currents[23, 24].

The possible chemical reactions occurring onto aluminum surfaces coated with PPy during the polarizations experiments are represented bellow[25].

Anodic reactions

$$Al \rightarrow Al^{+3} + 3e^-$$

(1)

Overoxidation of polymer

$$PPy_{undoped} \rightarrow PPy_{doped} + ne^-$$

(2)

Cathodic reactions

$$2 H_2O + O_2 + 4e^- \rightarrow 4 OH^-$$

(3)

$$PPy_{doped} + ne^- \rightarrow PPy_{undoped}$$

(4)

The electrochemical parameters obtained from the polarization curves shown in Figure 6 are presented in Table 1.

Table 1: Electrochemical parameters obtained for aluminum surfaces, just polished and coated with PPy films electrodeposited from aqueous solution containing each aliphatic acid, exposed to a 0.1 mol.L⁻¹ NaCl aqueous solutions

Aluminum surfaces	Time to reach the OCP in the chloride medium (h)	Ecorr(V/ Ag,AgCl,Cr)	jcorr(mA. cm-2)
Uncoated	48	-0.93	0.0023
Coated with PPy deposited in tartaric acid	24	-0.41	0.018
Coated with PPy deposited in oxalic acid	24	-0.46	0.024
Coated with PPy deposited in citric acid	24	-0.69	0.014

The corrosion current densities of the aluminum surfaces coated with PPy films were higher than those observed for uncoated surfaces. The occurrence of redox reactions in the polymeric matrix can also contribute to increasing the corrosion current densities values.

However, if the anodic current densities are used as criterion to determine the corrosion protection afforded by the PPy films, it would be possible to affirm that aluminum surfaces coated with the polymeric films are less susceptible to corrosion processes than just polished ones. Additionally, one could say that PPy films electrodeposited from tartaric acid offer better corrosion performance. This result is ascribed to the highest homogeneity of these films (SEM micrographs).

The presence of the voids among the *cauliflower* structure of the films deposited in oxalic and citric acid (Figure 5) allows the penetration of chloride ions (aggressive species) favoring the corrosion process.

It was also observed that the aluminum surfaces coated by PPy films deposited in tartaric acid presented smaller amount of pits after polarization experiments in chloride medium.

CONCLUSIONS

Considering the anodic current densities in the potentiodynamic polarization curves as a criterion to determining the corrosion protection afforded by the PPy films, it would be possible to affirm that aluminum surfaces coated with the polymeric films are less susceptible to corrosion process than just polished ones. Additionally, it could be also said that PPy films electrodeposited in tartaric acid medium are the ones that offer better corrosion performance. This result can be explained by the highest homogeneity of the polymeric films formed in this medium (SEM micrographs).

The PPy films deposited from oxalic or citric acid were more susceptible to the overoxidation process than those films formed in tartaric acid. This fact was demonstrated in the SEM micrographs and by the presence of the bands attributed, respectively, to bipolaronic species and to carbonyl groups on the FTIR spectra.

The overoxidation process is responsible by pores and defects along the polymeric structure that allow the penetration of chloride ions (aggressive species) favoring the corrosion process.

ACKNOWLEDGMENTS

The authors thank the Fundação de Amparo à Pesquisa do Estado de São Paulo (FAPESP) for the financial support and the Alcoa Alumínio SA, São Luis, Maranhão, Brazil.

REFERENCES

1.   Ingram MD, Staesche H, Ryder KS Activated polypyrrole electrodes for high power supercapacitor applications. *Solid State Ionics* 2004; 169(1):51-57.

2.   Ameer Q, Adeloju SB. Polypyrrole-based eletronic noses for environment and industrial analysis. *Sens. Actuators B.* 2005; 106(2):541-552.

3.   Tallman DE, Sokins G, Dominis A, Wallace GG. Electroactive conducting polymers for corrosion control. *J. Solid State Electrochem* 2002; 6:73-84.

4.   Tramontina J, Machado G, Azambuja DS, Pianicki CMS, Samios D. Removal of Cd^{+2} from aqueous solutions onto polypyrrole coated reticulated vitreous carbon electrodes. *Mat. Res.* 2001; 4(3):95-200.

5.   Wencheng S, Iroh JO. Effects of electrochemical process parameters on the synthesis and properties of polypyrrole coatings on steel *Synt Met* 1998; 95(3):159-167.

6.   Wang LX, Li XG, Yang YL. Preparation, properties and applications of polypyrrole *React. Funct Polymers* 2001; 47(2):125-139.

7.   Naoi K, Takeda M, Kanno H, Sakakura M, Shimada A. Simultaneous electrochemical formation of Al_2O_3 / polypyrrole layers: effect of electrolyte anion formation process. *Electrochim. Acta.* 2000; 45(20):3413-3421.

8.   Mohammad F. Comparative studies on diffusion behavior of electrochemically prepared polythiophene and polypyrrole: effect of ionic size of dopant. *Synth Met* 1999; 99(2):149-154.

9.   Kaplin DA, Qutubuddin S. Electrochemical synthetized polypyrrole films: effect of polymerization potential and electrolyte type. *Polymer* 1995; 36(6):1275-1286.

10.   Liu AS, Oliveira MAS. Electrodeposition of polypyrrole films on aluminum surfaces from tartrate aqueous solution. *J. Braz.Chem. Soc.* 2007; 18(1):143-152.

11.   ASTM G 102-89, "*Practice for Calculation of Corrosion Rates and Related Information from Electrochemical Measurements*"; 1994.

12. Saidman SB, Bessone JB. Electrochemical preparation and characterization of polypyrrole on aluminum in aqueous solution *J. Electroanal Chem.* 2002; 521(1):87-94.

13. Tsai ML, Chen PJ, Do JS. Preparation and characterization of PPy / Al_2O_3 / Al used as solid-state capacitor.*J. Power Sources* 2004; 133(2):302-311.

14. Mazhar AA, Arab ST, Noor EA. Electrochemical behavior of Al-Si alloys in acid alkaline media. *Bull. Electrochem* 2001; 17:449-458.

15. Bessone JB, Salinas DR, Mayer CE, Ebert M, Lorenz WJ an EIS study of aluminum barrier-type oxide films formed in different media. *Electrochim. Acta.* 1992; 37(12):2283-2290.

16. Muller B. Citric acid as corrosion inhibitor for aluminum pigment. *Corros. Sci.* 2004; 46(1):159-167.

17. Mazeikien R, Malinauskas A. Kinetics of the electrochemical degradation of polypyrrole. *Polym. Degrad. Stab.* 2002; 75(2):255-258.

18. Rodriguez I, Scharifker BR, Mostany J In situ FTIR study of redox and overoxidation process in polypyrrole films *J Electroanal. Chem.* 2000; 491(1):117-125.

19. Li Y, Qian R. Electrochemical overoxidation of conducting polypyrrole nitrate film in aqueous solutions.*Electrochim. Acta.* 2000; 45(11):1727-1731.

20. Marchewka MK, Debrus S, Pietraszko A, Barnes AJ, Ratajczak H. Crystal structure, vibration spectra and nonlinear optical properties of L-tartrate. *J. Molecular Structure.* 2003; 656(1):265-273.

21. Hon YM, Fung KZ, Lin SP, Hon MH Effect of metal ion sources on synthesis and electrochemical performance of spinel $LiMn_2O$ using tartaric acid *J. Solid State Chem.* 2002; 163(1):231-238.

22. Bazzaoui M, Martins L, Bazzaoui EA, Martins JI. New single step electrosynthesis process of homogeneous and strongly adherent polypyrrole films on iron electrodes in aqueous medium. *Electrochim. Acta* 2002; 47(18):2953-2962.

23. Breslin CB, Fenelon AM, Conroy KG. Surface engineering: corrosion protection using conducting polymers.*Materials and Design.* 2005; 26(3):233-237.

24.   Vilca DH, Moraes SR, Motheo AJ. Anodic treatment of aluminum in nitric acid containing aniline, previous to deposition of polyaniline and its role on corrosion. *Synth. Met* 2004; 140(1):23-27.

25.   Ocon P, Cristobal AB, Herrasti P, Fatas E. Corrosion performance of conducting polymer coating applied on mild steel. *Corros Sci.* 2005; 47(3):649-662.

Laboratory Assessment of Select Methods of Corrosion Control and Repair in Reinforced Concrete Bridges

Matthew D. Pritzl[1], Habib Tabatabai[2],
and Al Ghorbanpoor[2]

[1]Donan Engineering Co., Inc., 11321 Plantside Drive, Louisville, KY 40299, USA

[2]University of Wisconsin-Milwaukee, Department of Civil Engineering & Mechanics, 3200 North Cramer Street, Milwaukee, WI 53211, USA

ABSTRACT

Fourteen reinforced concrete laboratory test specimens were used to evaluate a number of corrosion control (CoC) procedures to prolong the life of patch repairs in corrosion-damaged reinforced concrete. These specimens included layered mixed-in chlorides to represent chloride contamination due to deicing salts. All specimens were exposed to accelerated corrosion testing for three months, subjected to patch repairs with various treatments, and further subjected to additional three months of exposure to accelerated corrosion. The use of thermal sprayed zinc, galvanic embedded anodes, epoxy/polyurethane coating, acrylic coating, and an epoxy patch repair material was evaluated individually or in combination. The specimens were assessed with respect to corrosion currents (estimated mass loss), chloride ingress, surface rust staining, and corrosion of the reinforcing steel observed after dissection. Results indicated that when used in patch repair applications, the embedded galvanic anode with top surface coating, galvanic thermal sprayed zinc, and galvanic thermal sprayed zinc with surface coating were more effective in controlling corrosion than the other treatments tested.

INTRODUCTION

Penetration of chlorides from deicing salts used on bridges causes significant long-term deterioration, which requires periodic maintenance and repair [1–3]. Corrosion of the reinforcing steel in concrete can lead to cracking and spalling concrete. Patch repairs are commonly used to address this problem. However, even when proper repair procedures are followed, failure of patches occurs in as little as 2–5 years [4].

The objective of this study was to assess relative performance of a number of corrosion control (CoC) procedures on patched chloride-contaminated reinforced concrete specimens tested under accelerated corrosion exposure in the laboratory. Fourteen specimens were exposed to accelerated corrosion testing for three months, subjected to patch repairs with various corrosion control treatments, and further subjected to an additional three months of exposure. The use of thermal sprayed zinc, galvanic embedded anodes, epoxy/polyurethane coating, acrylic coating, and an epoxy patch repair material was evaluated individually or in combination. The specimens were assessed with respect to corrosion currents (estimated mass loss), chloride ingress, surface rust staining, and corrosion of the reinforcing steel observed after dissection.

The phenomenon typically associated with patch failures, known as patch accelerated corrosion, occurs when the once "sound" area that surrounds the initial patch repair requires repair itself [4]. When traditional "chip and patch" repair procedures are used, a sudden change is introduced in the concrete surrounding the reinforcing steel as the bar crosses from old to new concrete (Figures 1 and 2). This occurs when new concrete (patch material), which is typically chloride-free and has a high pH, is placed adjacent to existing concrete, which is chloride-contaminated and has a lower pH. The interface creates zones of significantly different corrosion potentials along the steel bar. According to Ball and Whitmore, this difference in corrosion potential causes formation of new corrosion sites in the surrounding chloride-contaminated concrete [4].

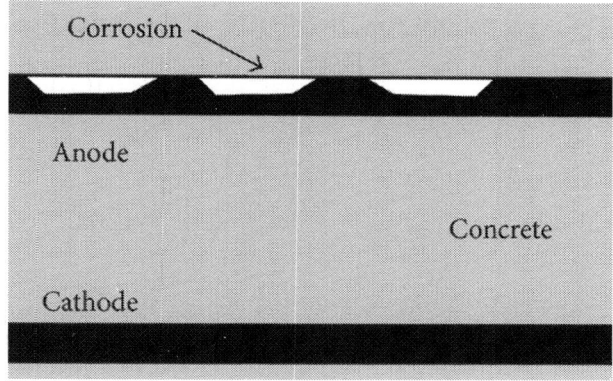

Figure 1: Corrosion cell in concrete.

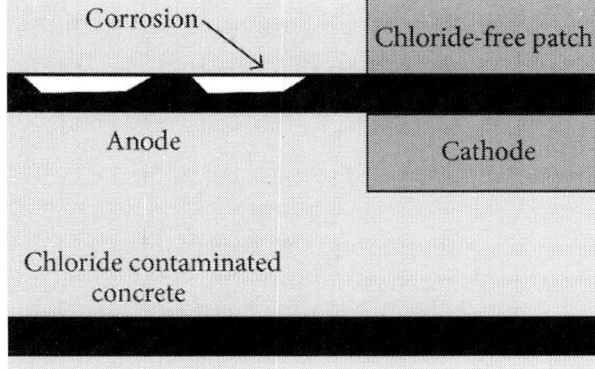

Figure 2: Patch-accelerated corrosion in concrete.

The patch-accelerated corrosion process (also referred to as the "ring anode" or "halo effect") occurs as the new concrete acts as an accelerant for corrosion. The steel bar within the new concrete can act as a cathode while the bar in the existing concrete acts as an anode. Evidence of this can be seen when spalls appear next to previously completed patch repairs [4].

In order to control further corrosion induced deterioration, strategies such as the use of concrete coatings, sealers, surfaced applied corrosion inhibitors, and cathodic protection have been used. Although coatings and sealers may work well for corrosion prevention, Tabatabai et al. found that surface applied treatments offered limited effectiveness when applied after the onset of corrosion [5].

Surface applied corrosion inhibitors, though not tested in this study, are applied on the hardened concrete surface and are intended to migrate through the pores of the concrete to protect the reinforcing bars from the ingress of harmful chemicals. El-Hacha et al. investigated the use of several surface applied corrosion inhibitors and concluded that the products tested "...generally helped delay and slow the corrosion process at the beginning, but none appeared to have totally stopped the corrosion process." El-Hacha et al. also concluded that "corrosion inhibitors seemed more effective at lower levels of chloride contamination and up to a threshold of about 0.5% by weight of cement" [6].

Cathodic protection (CP) makes use of an externally applied potential to shift all of the reinforcing steel into a cathodic and protected state. A 2001 Federal Highway Administration (FHWA) report concluded that "CP is only rehabilitation technique that has been proven to stop corrosion is salt-contaminated bridge decks regardless of the chloride content of the concrete" [7].

Cathodic protection systems are divided into two categories: impressed current cathodic protection (ICCP) and galvanic cathodic protection (GCP). Discrete anode GCP systems (which are embedded within concrete) are intended to provide local protection and have been used in patch repairs to avoid the "ring anode" or "halo" effect. These discrete sacrificial anodes are attached to the reinforcing bars within the patch area. When attached to the newly cleaned steel within the patched area, the sacrificial anode, being more electronegative, will corrode in preference to the steel in the adjacent, nonpatched area (Figure 3). Because of this, the "ring anode" or "halo" effect is supposed to be mitigated [4].

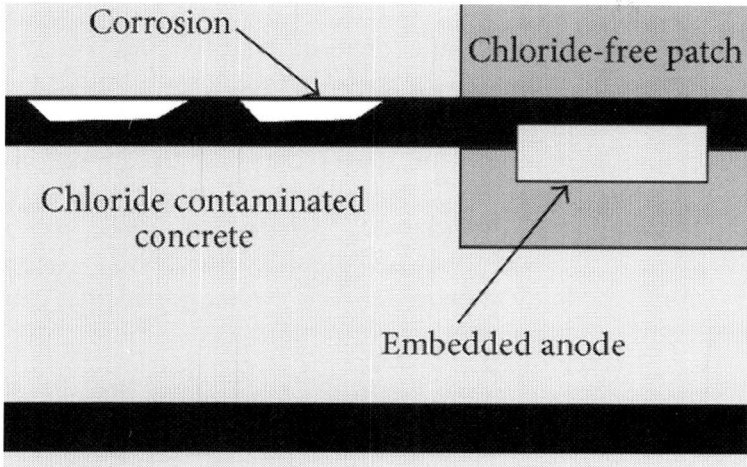

Figure 3: Intended action of discrete anode GCP in mitigating patch-accelerated corrosion.

A 2005 report by McMahan evaluated embedded galvanic units that were installed in the field in Vermont and were monitored for approximately 2 years. The report concluded that there were "…wide differences in monitored current, and presumably in corrosion rates and the amount of protection provided" and that the devices will "… only provide significant protection to concrete for 5 to 7 years" [8]. A 2007 paper by Dugarte et al. evaluated two types of commercial galvanic discrete anodes for reinforced concrete in both the field and laboratory. Based on preliminary findings, they concluded that "…only modest performance may be achieved with typical expected anode placement spacing in commonly encountered applications" [9].

Metalizing or thermal spraying is a method where a metal is melted and sprayed onto a prepared concrete surface. An electrical connection is then made between the embedded reinforcing steel and the sprayed metal. For reinforced concrete structures, the most commonly used thermal sprayed anodes are pure zinc and an aluminum-zinc-indium alloy (Al-Zn-In).

Daily and Green reported that the Al-Zn-In alloy will deliver more current than pure zinc in high resistivity environment (i.e., dry environments) because the indium keeps the anode more active [10]. However, Holcomb et al. suggested that a humectant can be added to pure zinc to increase moisture content at the zinc-concrete interface, thereby reducing the resistivity and increasing current output [11].

In 2003, Whitney et al. studied the use of cathodic protection on substructure elements in the splash zone for the Queen Isabella Causeway in Texas. The report stated that the galvanic sprayed zinc and aluminum-zinc alloy systems "…both performed reasonably well." The report also noted that although the zinc was less expensive, the aluminum-zinc alloy appeared to perform more effectively in dryer conditions and provided more uniform protection [12].

An accelerated corrosion testing approach involving concurrent use of the impressed current technique and salt water exposure has been successfully implemented by a number of researchers. Examples include works by Ray et al. [13], Mullard and Stewart [14], Michel et al. [15], and Abaosara et al. [16].

El Maaddawy and Soudki studied accelerated corrosion testing of reinforcing steel in concrete by varying impressed current densities. The authors noted that, "up to 7.27% mass loss, accelerated corrosion using the impressed current technique was effective in inducing corrosion of the steel reinforcement in concrete" [17]. Austin et al. also studied the electrochemical behavior of steel-reinforced concrete during accelerated corrosion testing and concluded that "the impressed current technique has been confirmed to be an effective and quick method of accelerating chloride-induced corrosion" [18].

MATERIALS AND METHODS

In this research, two (2) types of embedded discrete galvanic anodes, a humectant activated galvanic thermal sprayed zinc, a humectant activated galvanic thermal sprayed zinc with an epoxy/polyurethane coating, an embedded anode with an acrylic coating, a conventional cementitious patch repair material, and an epoxy patch repair material were chosen for evaluation. The thermal sprayed zinc and the various coatings were applied on the top surface of the concrete specimens.

The epoxy/polyurethane coating involved a first coat of epoxy followed by a second coat of polyurethane.

Specimens and Materials

Fourteen reinforced concrete test specimens were cast (Figure 4). Ready-mixed air-entrained concrete, containing fly ash, was used to fabricate the test specimens. The concrete was specified to meet the governing specifications for bridge deck construction in Wisconsin with a specified minimum 28th day compressive strength of 4,000 psi (27.6 MPa). The reinforced concrete test specimens had dimensions of 28 in. (71.1 cm) × 28 in. (71.1 cm) × 8 in. (20.3 cm). Number 5 reinforcing bars (diameter of 5/8 in. or 15.9 mm) meeting the requirements of ASTM A615 M were used as shown in Figure 5. Curing consisted of covering the specimens with plastic sheathing for seven days. The average measured 28-day compressive strength of three concrete cylinders was 5,839 psi (40.3 MPa).

Figure 4: Experimental laboratory setup.

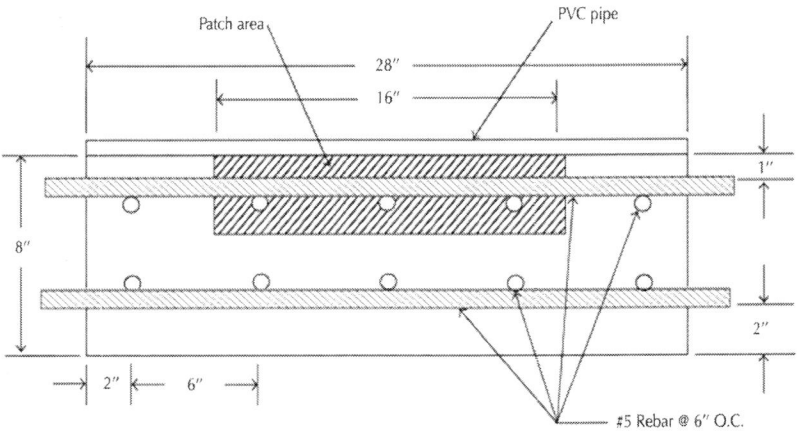

Figure 5: Cross-section of concrete specimens.

These 14 specimens (numbered 17 through 30) were part of a total of 30 such specimens that included 16 additional specimens for a companion study. To reduce the length of time needed for chlorides to reach the top steel layer, the top layer of reinforcement utilized a 1 in. (2.5 cm) clear cover. A standard 2 in. (5.1 cm) clear cover was used for the bottom layer of reinforcement. PVC pipe was caulked to the perimeter of the top surface of the concrete specimens to create the reservoir that periodically held the salt laden water (Figures 4 and 5). To better replicate chloride contaminated concrete, these 14 specimens were cast with layers of premixed chlorides. The bottom 5 in. (12.7 cm) of the specimens was cast without added chlorides while the upper 3 in. (7.6 cm) was cast with chloride profiles representative of common bridge deck conditions in the northern deicing states at a bridge age of 10 years.

Using chloride diffusion coefficient (D) and surface chloride concentration (C_0) from a Strategic Highway Research Program (SHRP) paper by Weyers et al. [19] and Fick's 2nd law of diffusion, a chloride profile that represented 10 years of exposure to chlorides was utilized. Based on the results of the SHRP study, a "D" of 0.11 in²/yr (0.71 cm²/yr) and a "C_0" of 5.985 lb/yd³ (3.55 kg/m³, representative of the mean of all the collected data) were used to determine the level of chlorides to be added in each of the three top layers of concrete.

Table 1: Level of chlorides mixed into the CoC specimens (1 inch = 25.4 mm and 1.0 lb/yd³ = 0.59 kg/m³)

Depth (inch)	Average depth (inch)	% chlorides by mass of concrete	Chloride content by volume of concrete
0-1	0.5"	0.113%	4.41 lb/yd³
1-2	1.5"	0.048%	1.87 lb/yd³
2-3	2.5"	0.014%	0.55 lb/yd³

At an average depth of 1.5 in. (3.8 cm), which is at the level of reinforcing steel in this project, the chloride content was nearly two times the corrosion threshold of 1.0 to 1.5 lb/yd³ (0.59 kg/m³ to 0.89 kg/m³) indicated by ACI 222 [20].

During the concrete pour, the bottom 5 inches (12.7 cm) of the CoC specimens was first placed. The three chloride profile levels were then added in succession. Concrete was mixed with table salt in a concrete mixer. When the pour was completed, all of the specimens were covered with a sheet of plastic for seven days.

The specimens were subjected to three months of exposure to accelerated corrosion (described later). After 3 months of exposure, the specimens were examined and tested for chloride ingress. Patch repairs were then performed (Figure 6) and the various treatments were applied. The patch repair consisted of saw-cutting the perimeter of the square patch area (16 in. (40.6 cm) × 16 in. (40.6 cm)), chipping out the concrete, cleaning the exposed reinforcing steel bar with a drill and wire wheel accessory, applying an epoxy coating to the steel bar, applying a bonding agent to the concrete substrate, and placing the patch repair material. The coating of the steel bar and the application of bonding agent within the patch area were based on patch material manufacturer's instructions and in accordance with conventional patch repair practices. Following anode installation directions, care was taken not to coat the points of electrical continuity (locations where the bars intersect), the connections of the anodes to the reinforcing steel, or the anodes themselves.

Figure 6: Patch repairs of laboratory specimens.

Table 2 describes the treatment(s) used on each specimen. The numbers on the specimens in Figure 4correspond to the specimen numbers in Table 2. While the embedded anodes were attached directly to the exposed bars, the thermal sprayed zinc and surface coatings were applied on the top surface of the concrete specimen and patched areas.

Table 2: Table of treatments used for each specimen*

Specimen number	Type of treatments	General description	Referred to as
17 and 18	Control	No treatment	Control
19 and 20	Thermal sprayed galvanic anode with coating applied on top surface of specimen	Humectant activated thermal sprayed zinc with epoxy/ polyurethane coating	TSZ w/ EP-C
21 and 22	Thermal sprayed galvanic anode applied on top surface of specimen	Humectant activated thermal sprayed zinc	TSZ

23 and 24	Embedded galvanic anode placed within patch area	Cylindrical-shaped zinc anode	EA-A
25 and 26	Embedded galvanic anode placed within patch area with acrylic coating applied on top surface of specimen	Cylindrical-shaped zinc anode with acrylic coating	EA-A w/A-C
27 and 28	Embedded galvanic anode placed within patch area	Box-shaped zinc anode	EA-B
29 and 30	Epoxy repair mortar as a patch material	Epoxy resins and polyamino amine adducts	EM

Specimens 17 through 28 included a conventional cement-based patch material in addition to the treatments shown. Specimens 29 and 30 included an epoxy patch material. All patching and treatments were applied after 3 months of exposure to accelerated corrosion.

Two types of patch materials were used: a conventional material and a proprietary epoxy-based material. The conventional patch repair material used on specimens 17 through 28 was a commercial cement-based, rapid strength gain, patching and repair mortar which contained a migratory corrosion inhibitor. According to the manufacturer, the patch material is compatible with galvanic anodes. The epoxy patch repair material utilized on specimens 29 and 30 is reported by the manufacturer to be a three component, solvent-free, high performance epoxy mortar.

Epoxy bonding agents for coating of substrate concrete (within the patch area) were recommended by the manufacturer of the conventional patches. Yet, epoxy bonding agents are generally not recommended for use with galvanic anodes. However, epoxy bonding agents can be used with embedded anodes if both the metallic and ionic paths are maintained. Since the metallic path had already been confirmed (through the connection between the anode and the bars), the ionic path from the anode to the cathode had to be provided as well. For the discrete anodes, this was accomplished by not coating the anodes or the substrate concrete immediately below the anodes. When using epoxy bonding agents in specimens that received thermal sprayed metals, the ionic path will still reach the bars in the areas inside and outside of the patch with direct contact to the top surface of each specimen.

The manufacturer of the conventional patch material recommended the use of an epoxy bonding agent and the manufacturer of the epoxy patch repair material specified a concrete primer on the surface of the concrete substrate within the patch area. We chose to use an epoxy bonding agent in conjunction with the conventional patch repair materials in accordance with manufacturer's recommendation, following all guidelines applicable for concurrent use of embedded anodes. Moreover, all guidelines for the epoxy patch repair material were followed.

Experimental Methods

The treatments in question were evaluated with respect to corrosion currents, chloride ingress, extent of cracking, severity of rust staining, and visual inspection of the reinforcing steel after the conclusion of testing and dissection.

To accelerate the corrosion process, the specimens were subjected to wetting/drying cycles and a reverse cathodic protection system. Cycles of one week wet (using a 6% NaCl solution on the top surface) and one week dry (no saltwater ponding) were utilized. A reverse cathodic protection system was created by continuously applying a regulated voltage of 9 V from the positive terminal of the regulated power supply to the top layer of reinforcement (the anode). A 1 Ω precision resistor located between the positive terminal and the anode was used to facilitate measurement of current (Figure 7). The voltage across the 1 Ω resistor was manually recorded each day on all specimens using a high impedance multimeter.

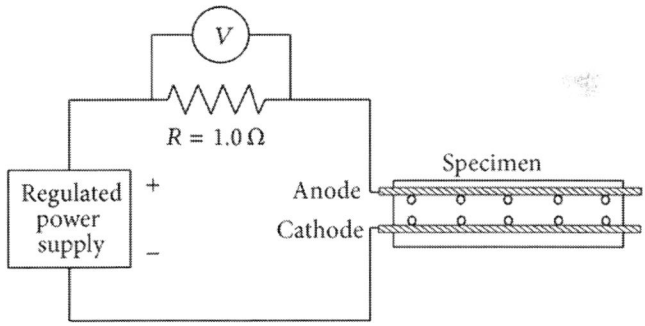

Figure 7: Corrosion cell for laboratory specimens.

The measured corrosion currents were used to estimate the amount of steel loss that had occurred for comparison purposes. The amount of steel loss was estimated using Faraday's Law (1),

$$m = \frac{A^{tm}C}{Fz},$$

(1)

where m = loss of mass, A^{tm} = atomic mass of the reaction ion (55.85 g/mol for iron), C = total charge that has passed through the circuit = $\int I(t)dt$, $I(t)$ = measured corrosion current at time (t), F = Faraday's constant (96485 C/mol), and Z = valence of reaction (assumed to be 2).

The chloride content of all specimens was determined by analyzing drilled concrete samples at various depths using the Rapid Chloride Test (RCT) method [21]. Acid soluble chloride levels were measured as a percentage of concrete mass. Regression analyses utilizing Fick's 2nd Law (2) and Microsoft Excel's solver function were utilized to further analyze the chloride content results and to provide a direct comparison among the specimens,

$$C_{(x,t)} = C_0 \left(1 - \text{erf} \frac{x}{2\sqrt{Dt}} \right),$$

(2)

$C^{(x,t)}$ = chloride concentration at depth x and time t, C^0 = surface chloride diffusion (lb/yd^3 or kg/m^3), erf = error function (a mathematical function), and D = chloride diffusion coefficient (in^2/yr or cm^2/yr).

Prior to patch repairs, the extent of rust staining seen on the concrete surface was evaluated visually and quantified. The exposed reinforcing steel was visually evaluated for extent of corrosion after the concrete in the patch area was removed.

Periodically, half-cell potential readings were taken. The readings were taken at sixteen locations per specimen. Furthermore, detailed crack maps were generated at 0-month, 3-month, and 6-month exposure. The widths of the cracks were measured using a standard crack width comparator. At the conclusion of testing, the extent of rust staining on the concrete surface was evaluated visually. Finally, the specimens were dissected and the embedded reinforcing steel was visually assessed for extent of corrosion.

The results of the various measurements (as discussed in Section 3) were given numerical indices (3) according to a scale of 0 (minimum) to 4 (maximum). The actual value is the value associated with the parameter of interest for each specimen, while the minimum and maximum values used in (3) are based on the minimum and maximum values observed across all specimens. For example, the minimum and maximum steel loss values across all specimens corresponded to ratings of 0 and 4, respectively,

$$\text{Index} = \left(\frac{(\text{actual value}) - (\text{minimum value})}{(\text{maximum value}) - (\text{minimum value})} \right) \times 4. \tag{3}$$

RESULTS AND DISCUSSION

A detailed report of this experimental program and its individual results is provided by Tabatabai et al. [22]. In the following discussion, the average results are shown for brevity. Individual results for companion specimens used in averaging were in reasonable agreement.

Corrosion Currents

A plot of average corrosion current versus time for the specimens is shown in Figure 8. As expected, the corrosion currents increased during the wet periods (shaded region) and decreased during the dry periods.

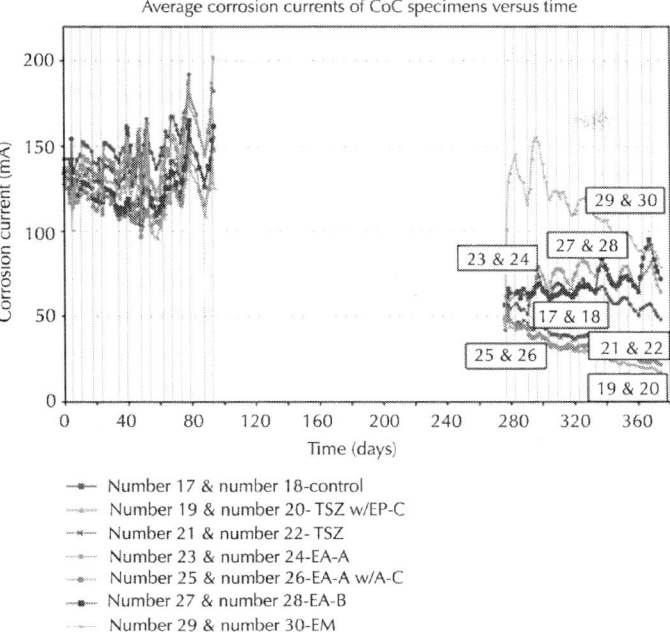

Figure 8: Average corrosion currents of specimens.

From the initiation of testing until 3 months (90 days), there appears to be reasonable agreement among all specimens (as expected). The break in the graph between 3 months (90 days) and the continuation of the project at 270 days was due to the time necessary to perform the chloride testing and patch repairs. Accelerated testing (including the 9 V electrical potential) was restored after 270 days and continued for another three months.

After patch repairs, the TSZ, EA-A w/A-C, and TSZ w/EP-C treatments all displayed a decrease in corrosion current, while the currents associated with the Control and EM specimens increased initially and then decreased. It is believed that the rapid increase in corrosion current for the EM was due to the widely dissimilar properties of the epoxy patch repair material compared with the surrounding concrete. A reinforcing bar crossing from the substrate concrete into the new epoxy material would likely develop more pronounced anode and cathode area on the bar. Meanwhile, EA-A and EA-B also exhibited an increase in corrosion current over time after treatment.

Steel Loss

By utilizing the aforementioned corrosion currents and (1), the amount of steel loss was estimated (Table 3). Numerical integration was used to calculate the total charge used in estimating steel loss.

Table 3: Steel loss of CoC specimens after 3-month exposure and 6-month exposure

Specimen number	Treatment	0–3- month steel loss (g)	3–6- month steel loss (g)	3–6- month index	0–6- month steel loss (g)	0–6- month index
17 and 18	Control	357.2	148.7	1.3	505.9	2.8
19 and 20	TSZ w/EP-C	336.4	73.6	0.1	410.0	1.2
21 and 22	TSZ	313.3	90.6	0.4	403.9	1.1
23 and 24	EA-A	338.1	175.8	1.8	513.9	2.9
25 and 26	EA-A w/A-C	278.2	78.3	0.2	356.5	0.3
27 and 28	EA-B	295.4	169.2	1.7	464.6	2.1
29 and 30	EM	289.0	275.4	3.4	564.4	3.8

For the 0–3 month data, the average steel loss was calculated to be 315.4 g (0.70 lb) with a standard deviation of 35.4 g (0.08 lb). Based on the initial steel loss values, it appears that all specimens were in a reasonably similar condition after the first 3 months of laboratory testing.

For the 3–6 month steel loss data, the TSZ w/EP-C, EA-A w/A-C, and TSZ produced the lowest indices. When considering the 0–6 month index, EA-A w/A-C, TSZ, and TSZ w/EP-C had the lowest Index values. Therefore, it can be concluded that these three treatments performed better with regard to theoretical steel loss due to corrosion. It is interesting to note that all these better-performing treatments incorporated some form of coating or physical barrier on the top surface of the specimen.

Chloride Ingress

The baseline chloride content was taken from virgin concrete. The average measured chloride content of the virgin concrete was found to be 0.042% by concrete weight or approximately 1.65 lb/yd³ (0.98 kg/

m^3) of concrete. This measured chloride content was relatively high. An earlier study by Tabatabai et al. has similar results and found that the source of the chlorides was from coarse limestone aggregates [5].

Prior to exposure to accelerated corrosion, the chloride contents of all 14 specimens were evaluated at average depths of 1/4 in. (0.64 cm), 1/2 in. (1.27 cm), 3/4 in. (1.91 cm), 1 in. (2.54 cm), 1.25 in. (3.18 cm), 1.5 in. (3.81 cm), 2 in. (5.08 cm), 2.5 in. (6.35 cm), and 3 in. (7.62 cm), so that confirmation of the actual mixed-in chloride contents could be made. Three locations, for a total of 27 chloride tests per specimen, were analyzed. Chloride testing (Table 4) revealed that the actual chloride contents were in reasonable agreement with the "initial intended plus baseline chloride" profile.

Table 4: Comparison of initial intended and average acid-soluble chloride contents of CoC specimens (% chlorides by concrete weight) at 0 month (1 inch = 25.4 mm)

Depth	Initial intended plus baseline chlorides	Average of initial measured chlorides	Average initial measured chlorides
0" to $\frac{1"}{4}$	0.155	0.183	0.156
$\frac{1"}{4}$ to $\frac{1"}{2}$		0.171	
$\frac{1"}{2}$ to $\frac{3"}{4}$		0.142	
$\frac{3"}{4}$ to 1"		0.129	

1" to 1$\frac{1}{4}$"	0.090	0.119	0.106
1$\frac{1}{4}$" to 1$\frac{1}{2}$"		0.109	
1$\frac{1}{2}$" to 2$\frac{1}{2}$"		0.089	
2" to 2$\frac{1}{2}$"	0.056	0.073	0.065
2$\frac{1}{2}$" to 3$\frac{1}{2}$"		0.057	

By utilizing regression analyses of measured chlorides, the agreement between the "initial intended" and "initial measured" chlorides could be further verified. Using a time of 10 years (assumed for calculating the amount of mixed-in chlorides), C_0 was found to equal 0.149% chlorides by concrete weight (5.83 lb/yd^3 or 3.46 kg/m^3) and D_{avg} was found to equal 0.150 in^2/yr (0.97 cm^2/yr), with a standard deviation of 0.026 in^2/yr (0.17 cm^2/yr). These values are in reasonable agreement with the values of C_0 = 0.153% (6.0 lb/yd^3 or 3.56 kg/m^3) and D = 0.110 in^2/yr (0.71 cm^2/yr) that were used initially to determine the mixed-in chloride levels.

Prior to removing concrete from the specimens for patch repair, the specimens were again evaluated for chlorides after 3 months of accelerated corrosion testing. Testing was performed at 1/4 in. (0.64 cm) intervals to a depth of 2 in. (5.08 cm) at 3 locations per specimen. Table 5 compares the average measured chloride contents at 0 and 3 months.

Table 5: Comparison of average acid-soluble chloride contents of CoC specimens at 0 month and after 3 months (1 inch = 25.4 mm)

Depth	Average of 0-month chlorides	Average of 0-month chlorides per inch	Average of 3-month chlorides	Average of 3-month chlorides per inch
0" to $\frac{1''}{4}$	0.183	0.156	0.502	0.322
$\frac{1''}{4}$ to $\frac{1''}{2}$	0.171		0.367	
$\frac{1''}{2}$ to $\frac{3''}{4}$	0.142		0.243	
$\frac{3''}{4}$ to 1"	0.129		0.174	
1" to $1\frac{1''}{4}$	0.119		0.131	0.108
$1\frac{1''}{4}$ to $1\frac{1''}{2}$	0.109	0.106	0.113	
$1\frac{1''}{2}$ to 2"	0.089		0.099	
2" to $2\frac{1''}{2}$			0.088	

From Table 5, it is clear that the chlorides were effectively drawn into top 1 in. (2.54 cm) of the concrete during the first 3 months of exposure. The average chloride content in the top 1-inch (2.54 cm) of concrete more than doubled in the 3-months of accelerated corrosion testing.

An optimization analysis using the "3-month measured chlorides," minus the base-line chlorides, with a time of 0.25 years (3 months), revealed that the following parameters of Fick's 2nd Law best fit the experimental data: C_0 = 0.514% by concrete weight (20.12 lb/yd³ or 11.9 kg/m³) and D_{avg} = 1.375 in²/yr (8.87 cm²/yr) with a standard deviation of 0.565 in²/yr (3.65 cm²/yr).

The agreement between the intended, measured, and calculated chlorides (based on regression) is shown in Figure 9 (for specimen number 17, a control specimen). This representative graph compares the measured concrete chlorides before the start of accelerated corrosion with the intended (by mixing salt) chloride profile plus the chloride level existing in the concrete itself (base-line chloride). This graph also compares the measured chlorides at 3 months with the projected chlorides at 3 months based on estimated D and C_0 values.

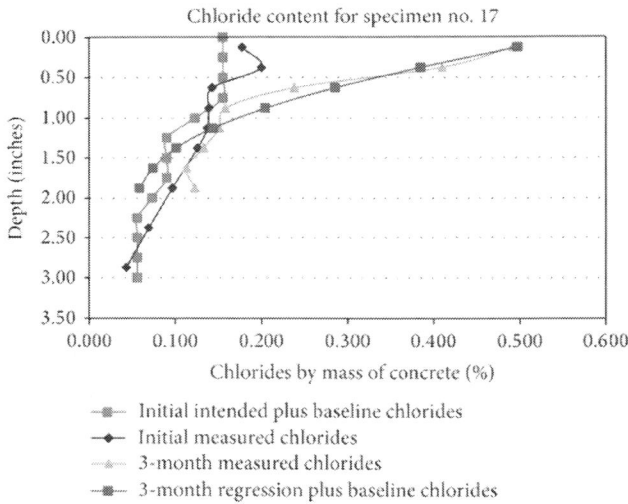

Figure 9: Comparison of initial and 3-month chlorides for specimen number 17.

Prior to patch repairs, the average base-line chloride content of the conventional repair material was found to be 0.008% by concrete weight. This level of chlorides is well within the accepted limits. The base-line chloride content of the EM was found to be 0.001% chlorides by weight.

After exposure to additional three months of accelerated corrosion testing, the CoC specimens were again tested for chloride ingress. Chloride testing was performed at locations in the original (substrate) concrete as well as in the patch area.

For the substrate concrete, the chloride profiles of most specimens did not agree with Fick's Law. Therefore, the use of regression in analyzing the substrate concrete after 6 months of exposure was not warranted. However, the chloride contents of the substrate concrete were not used in evaluating the performance of the treatments within the patch repair materials. Chloride testing of the substrate concrete was performed to show that chlorides continued to penetrate the concrete. It should be noted that specimens number 19 through 22 (those with thermal sprayed zinc) had nonconforming chloride levels at a depth of 1/4 in. (0.64 cm) only. It is believed that the TSZ (zinc anode) attracts and retains negatively charged chloride ions near the surface, thus causing deviation from Fick's 2nd Law.

Analysis of the chloride testing data for the patch repair materials after additional 3 months of testing revealed that, in general, chlorides were only drawn into the top 1/4 in. (0.64 cm) of the patch repair materials. Table 6shows the estimated C_0 and D values for the patch materials after 3 months of exposure based on an optimization analysis assuming that Fick's Law applies.

Table 6: Calculated chloride diffusion coefficients for patch materials of CoC specimens at the conclusion of project (1 inch = 25.4 mm)

Specimen number	Patch treatment		C_0 (% Cl by concrete mass)	$D_{Treatment}$ (in²/yr)	Index
17 and 18	Control	(Conventional repair material)	0.445	0.030	2.8
19 and 20	TSZ w/ EP-C			0.010	0.9
21 and 22	TSZ			0.004	0.4
23 and 24	EA-A			0.010	0.9
25 and 26	EA-A w/A-C			0.005	0.4
27 and 28	EA-B			0.021	1.9
29 and 30	EM			—	—

As displayed in Table 6, TSZ and EA-A w/A-C appeared to be the most effective in reducing the ingress of chlorides into the conventional patch repair materials. The specimens with epoxy mortar (EM) had low chloride contents; however, the distribution was not consistent with Fick's Law. Therefore, a "D" value is not reported for EM specimens.

Surface Staining and Steel Corrosion

To provide a quantitative measure of the condition of the specimens after exposure to accelerated corrosion testing, a visual examination of the rust staining on the concrete surface and exposed reinforcing steel was performed, so that a numerical rating could be assigned to each specimen. The condition of the CoC specimens after 6 months of exposure and dissection is shown in Figures 10 and 11, respectively.

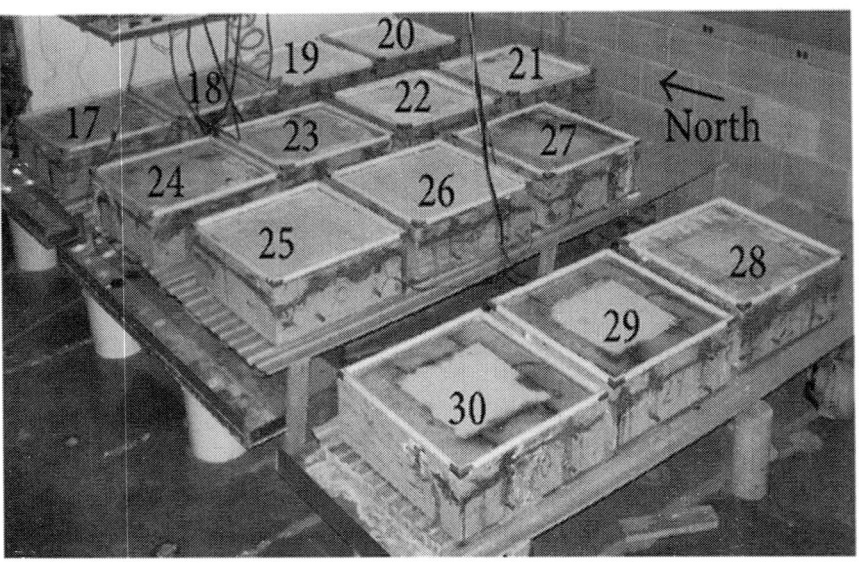

Figure 10: Surface staining on laboratory CoC specimens after 6 months.

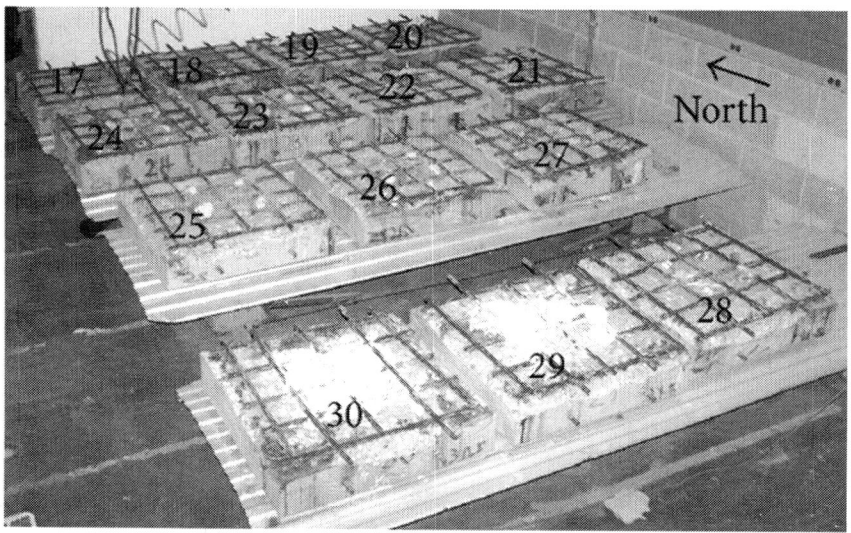

Figure 11: Dissected laboratory CoC specimens.

Based on a rating scale of 0 to 4, with 0 being the best condition and 4 being the worst condition, rust staining on the surface of the specimens as well as the level of section loss in the reinforcing steel was evaluated. The rating scale for staining was based purely on visual examination; the more severe the staining, the higher the grade. The rating scale for the condition of the reinforcing steel was based on the loss of ribs (Figures 12 and 13). If no corrosion by-products were present, a rating of 0 was given. If it appeared that all of the ribs were lost, a rating of 4 was given. If, on average, 1/4, 1/2, or 3/4 of the ribs were lost, ratings of 1, 2, and 3 were given, respectively. This rating was given to each of the top-layer bars in each specimen. The ratings were then averaged for each specimen. The two separate ratings (staining and loss of steel section) were then added together, for a maximum value of 8, to determine a combined rating.

Figure 12: Steel rating of 2 and 3 (close-up of exposed Rebar of control after 3 months).

Figure 13: Steel rating of 4 (close-up of exposed Rebar of TSZ w/EP-C after 6 months).

After the initial 3 months of exposure, it was found that the average rating for the exposed reinforcing steel (within the excavated patch area) was 3.0 and that the average rating for the staining was 3.0. The average 3-month total (steel bar plus staining) rating for all CoC specimens was 6.0.

Table 7 presents the individual and final ratings for each of the specimens after 6 months of exposure. In Table7, the five reinforcing bars (in the top layer in each specimen) are designated A through E. Rebar A is located at the west end of the specimens, Rebar B through Rebar D are located within the patch, and Rebar E is located on the east end of the specimens.

Table 7: Rating of concrete staining and reinforcing steel for CoC specimens after 6 months

Specimen number	Individual Rebar								"in" Rebar average	Surface staining	Total rating
	A	B		C		D		E			
	out	in	out	in	out	in	out	out			
17 and 18	4.0	3.0	4.0	3.0	3.5	3.0	4.0	4.0	3.0	0.5	3.5
19 and 20	3.0	3.5	4.0	3.0	3.5	4.0	4.0	4.0	3.5	0.0	3.5
21 and 22	4.0	3.5	4.0	3.0	4.0	4.0	4.0	3.5	3.5	0.0	3.5
23 and 24	4.0	4.0	4.0	2.5	3.0	4.0	4.0	4.0	3.5	1.0	4.5
25 and 26	3.5	4.0	4.0	3.0	4.0	4.0	4.0	3.0	3.7	0.0	3.7
27 and 28	3.0	3.5	4.0	2.5	3.5	4.0	4.0	4.0	3.3	2.0	5.3
29 and 30	3.5	3.0	4.0	3.0	4.0	3.5	4.0	4.0	3.2	3.0	6.2

All exposed bars were rated; however, only the Rebar that was used within the patch repairs ("in" bars) was counted toward the bar rating (bars B through D). Bars outside of the patch area ("out" bars) were not counted toward the rating of the steel rebar. In doing so, a comparison could be made with the ratings that were made prior to patch repairs. Concrete surface staining was also similarly rated.

Half-Cell Potential

Half-cell measurements utilizing a copper-copper sulfate electrode were obtained for each of the concrete specimens. Prior to measurement, the accelerated corrosion system was turned off for a day and the slabs were saturated with tap-water. Readings were made after 3 months. It was found that the average potential was −546.1 mV with a standard deviation of −28.4 after three months of exposure (Table 8). Because the readings were fairly uniform, contour plots were not made. Since these readings were more negative than −350 mV, a 90% probability of corrosion was indicated after 3 months of exposure in the CoC specimens [2]. Half-cell potential testing was not performed after 6 months of exposure because of the high potential levels found at 3 months in all specimens and because of a number of coatings applied on the top surface of specimens that would prevent half-cell potential measurements.

Table 8: Half-cell potential readings after 3 months

Specimen number	Future treatment	Average (mV)	Std Dev. (mV)
17 and 18	Control	−541.1	27.9
19 and 20	TSZ w/EP-C	−540.6	27.3
21 and 22	TSZ	−532.7	28.1
23 and 24	EA-A	−542.9	22.7
25 and 26	EA-A w/ A-C	−554.2	24.9
27 and 28	EA-B	−556.8	32.3
29 and 30	EM	−554.3	18.2

Discussion

The following section provides a summary and discussion of the results of this study. Based on chloride content, reinforcing steel corrosion, concrete surface staining, and half-cell potential testing, the CoC specimens appeared to be in a similar condition after the first 3 months of accelerated corrosion testing. Therefore, it can be concluded that the addition of chlorides to the concrete mix, "ponding" of salt water, and application of electric current were controlled properly during the first 3 months of exposure and all specimens (similarly prepared and exposed for 3 months) were responding similarly.

A summary of numerical ratings for the CoC specimens after 6 months of exposure is presented in Table 9. For this evaluation, the chloride content of the substrate concrete was not used in evaluating the effectiveness of the treatments within the patch repair and only the "in" patch reinforcing steel, along with surface staining (including substrate concrete), was used for the rating.

Table 9: Condition summary of CoC specimens after 6 months of exposure

Specimen number	Treatment	6-month CoC ratings			Total (out of 16)
		Steel loss (out of 4)	Patch chloride content (out of 4)	Rebar corrosion and staining (out of 8)	
17 and 18	Control	2.80	2.80	3.50	9.10
19 and 20	TSZ w/ EP-C	1.25	0.90	3.50	5.65
21 and 22	TSZ	1.10	0.35	3.50	4.95
23 and 24	EA-A	2.90	0.90	4.50	8.30
25 and 26	EA-A w/A-C	0.35	0.40	3.67	4.42
27 and 28	EA-B	2.10	1.95	5.33	9.38
29 and 30	EM	3.75	—*	6.17	—*

Patch chloride content did not conform to Fick's 2nd Law.

Based on the results of Table 9, EA-A w/A-C, TSZ, and TSZ w/ EP-C appeared to be the most effective in controlling corrosion. It is important to note that all these treatments incorporated some form of coating or physical barrier applied on the top surface of specimens; however, none of the CoC specimens had a coating treatment alone to conclusively assess whether the coating is the sole determining factor.

When evaluating the patched areas of the TSZ and TSZ w/EP-C

specimens, it is interesting to note that the steel bar near the connection point to the thermal sprayed zinc exhibited more corrosion than other areas of the steel bar.

The patches containing the EA-A (without coating) and EA-B were no more effective in controlling corrosion than the control specimens. Chloride measurements indicated that the anodes attracted chlorides to their vicinity. These chloride "hot spots" increased corrosion of the reinforcing steel in the area of attachment to the anodes.

In regard to the EM specimens, the high initial corrosion currents and associated steel loss appear to be the result of the highly dissimilar material properties. A "ring-anode" effect became visible at the interface of the patch and substrate concrete over the duration of testing. Although the EM patch material itself did not display any signs of cracking, the existence of the "ring-anode" effect at the perimeter of the patch was clearly evident. These were the only specimens to significantly display this phenomenon.

It appeared that the conventional patch repair material performed well (i.e., patch itself did not deteriorate). This was evidenced by the fact that no significant cracking was present on any of the patched areas at the conclusion of testing. However, as stated earlier, the performance of the patch material is not an overall indication of performance of patch due to the "halo effect."

CONCLUSIONS

Based on the observation of the test specimens subjected to an accelerated corrosion regime, the following conclusions are made.(1) Embedded anode A with acrylic coating (EA-A w/A-C), thermal sprayed zinc (TSZ), and thermal sprayed zinc with epoxy/polyurethane coating (TSZ w/EP-C) were most effective in controlling corrosion. It appears that the main factor for the better performance of these specimens was the presence of some form of coating or physical barrier.(2)The performance of the embedded anodes (EA-A without coating as well as EA-B without coating) was not better than the control specimens.(3)For the epoxy patch repair material (EM), the initial increase in corrosion current and appearance of the "ring-anode" effect at the perimeter of the patch is due to the significant dissimilarity between epoxy mortar and concrete.

ACKNOWLEDGMENTS

The authors wish to express their gratitude and sincere appreciation to the Wisconsin Highway Research Program for funding this effort. The project team would also like to thank Ambassador Steel of Waukesha, Wisconsin, for donating materials, Masonry Restoration, Inc. of Milwaukee, Wisconsin, for lending their demolition equipment, Aaron Coenen, John Condon, Chin Wei-Lee, Cory Schultz, and Rahim Reshadi for assisting with the placement of concrete, and Dr. Tracy Pritzl for her assistance with chloride testing.

REFERENCES

1.   United States Department of Transportation Federal Highway Administration Turner-Fairbank Research Center, 1998, Corrosion Protection: Concrete Bridges, Turner-Fairbank Research Center,http://www.tfhrc.gov/structur/corros/corros.htm.

2.   P. H. Emmons, Concrete Repair and Maintenance Illustrated, RS Means, Kingston, Mass, USA, 1993.

3.   M. El-Reedy, Steel-Reinforced Concrete Structures: Assessment and Repair of Corrosion, Taylor & Francis Group, Boca Raton, Fla, USA, 2008.

4.   C. J. Ball and D. W. Whitmore, Corrosion Mitigation Systems For Concrete Structures, Concrete Repair Bulletin, 2003.

5.   H. Tabatabai, A. Ghorbanpoor, and A. Turnquist-Nass, Rehabilitation Techniques For Concrete Bridges, Wisconsin Department of Transportation Research and Library, Madison, Wis, USA, 2005.

6.   R. El-Hacha, A. Mirmiran, A. Cook, and S. Rizkalla, "Effectiveness of surface-applied corrosion inhibitors for concrete bridges," Journal of Materials in Civil Engineering, vol. 23, no. 3, pp. 271–280, 2011. · ·

7.   United States Department of Transportation Federal Highway Administration Research and Development, 2001, Long-Term Effectiveness of Cathodic Protection Systems on Highway Structures, McLean.

8.   J. McMahan, Preliminary Evaluation of Galvashield Installation in BR 48, I-89, Waterbury, Vermont Agency of Transportation, Montpelier, Vermont, 2005.

9.   M. Dugarte, A. A. Sagüés, R. Powers, and I. Lasa, "Evaluation of point anodes for corrosion prevention in reinforced concrete," Tech. Rep. 07304, NACE International, Houston, Tex, USA, 2007.

10.   S. F. Daily and W. K. Green, "Galvanic cathodic protection of reinforced and prestressed concrete structures using CORRYSPRAY—a thermally sprayed aluminum alloy," Tech. Rep., Corrpro Companies.

11.   G. R. Holcomb, B. S. Covino Jr., S. D. Cramer, et al., Humectants To Augment Current From Metallized Zinc Cathodic Protection Systems on Concrete, Oregon Department of Transportation, Salem, Mass, USA, 2002.

12.   D. Whitney, L. Elcheverry, and H. Wheat, Cathodic Protection: Coordinating Corrosion To Save State Structures, Center for Transporation Research: The University of Texas at Austin, Austin, Tex, USA, 2003.

13.   I. Ray, G. C. Parish, J. F. Davalos, and A. Chen, "Effect of concrete substrate repair methods for beams aged by accelerated corrosion and strengthened with CFRP," Journal of Aerospace Engineering, vol. 24, no. 2, pp. 227–239, 2011. · ·

14.   J. A. Mullard and M. G. Stewart, "Corrosion-induced cover cracking: new test data and predictive models," ACI Structural Journal, vol. 108, no. 1, pp. 71–79, 2011.

15.   A. Michel, B. J. Pease, M. R. Geiker, H. Stang, and J. F. Olesen, "Monitoring reinforcement corrosion and corrosion-induced cracking using non-destructive x-ray attenuation measurements," Cement and Concrete Research, vol. 41, no. 11, pp. 1085–1094, 2011. · ·

16.   L. Abosrra, A. F. Ashour, and M. Youseffi, "Corrosion of steel reinforcement in concrete of different compressive strengths," Construction and Building Materials, vol. 25, no. 10, pp. 3915–3925, 2011. · ·

17.   T. A. El Maaddawy and K. A. Soudki, "Effectiveness of impressed current technique to simulate corrosion of steel reinforcement in

concrete," Journal of Materials in Civil Engineering, vol. 15, no. 1, pp. 41–47, 2003. · ·

18. S. A. Austin, R. Lyons, and M. J. Ing, "Electrochemical behavior of steel-reinforced concrete during accelerated corrosion testing," Corrosion, vol. 60, no. 2, pp. 203–212, 2004.

19. R. E. Weyers, E. W. Weyers, et al., Concrete Bridge Protection and Rehabilitation: Chemical and Physical Techniques—Service Life Estimates, Strategic Highway Research Program, National Research Council, Washington, DC, USA, 1994.

20. American Concrete Institute Committee 222, ACI 222R-01: Protection of MetaLs in Concrete Against Corrosion, American Concrete Institute, 2001.

21. Germann Instruments, 2006, RCT & RCTW, Summary of Germann Instruments,http://www.germann.org/Brochures/RCT.pdf.

22. H. Tabatabai, A. Ghorbanpoor, and M. D. Pritzl, "Evaluation of select methods of corrosion prevention, corrosion control, and repair in reinforced concrete bridges," Final Report, Wisconsin Highway Research Program, Madison, Wis, USA.

3

Greener Approach towards Corrosion Inhibition

Neha Patni, Shruti Agarwal, and Pallav Shah

Department of Chemical Engineering, Institute of Technology, Nirma University, S. G. Highway, Ahmedabad, Gujarat 382481, India

ABSTRACT

Corrosion control of metals is technically, economically, environmentally, and aesthetically important. The best option is to use inhibitors for protecting metals and alloys against corrosion. As organic corrosion inhibitors are toxic in nature, so green inhibitors which are biodegradable, without any heavy metals and other toxic compounds, are promoted. Also plant products are inexpensive, renewable, and readily available. Tannins, organic amino acids, alkaloids, and organic dyes of plant origin have good corrosion-inhibiting abilities. Plant extracts contain many organic compounds, having polar atoms such as O, P, S, and N. These are adsorbed on the metal surface by these polar atoms, and protective films are formed, and various adsorption isotherms are obeyed. Various types of green inhibitors and their effect on different metals are mentioned in the paper.

INTRODUCTION

Corrosion is the deterioration of materials by chemical interaction with their environment. The term corrosion is sometimes also applied to the degradation of plastics, concrete, and wood, but generally refers to metals. The most widely used metal is iron (usually as steel). Corrosion can cause disastrous damage to metal and alloy structures causing economic consequences in terms of repair, replacement, product losses, safety, and environmental pollution. Due to these harmful effects, corrosion is an undesirable phenomenon that ought to be prevented [1]. There are several ways of preventing corrosion and the rates at which it can propagate with a view of improving the lifetime of metallic and alloy materials. The use of inhibitors for the control of corrosion of metals and alloys which are in contact with aggressive environment is one among the acceptable practices used to reduce and/or prevent corrosion. A corrosion inhibitor is a substance which, when added in small concentration to an environment, effectively reduces the corrosion rate of a metal exposed to that environment.

Corrosion inhibitors can be divided into two broad categories, namely, those that enhance the formation of a protective oxide film through an oxidizing effect and those that inhibit corrosion by selectively adsorbing on the metal surface and creating a barrier that prevents access of corrosive agents to the metal surface [1]. Almost all organic molecules containing heteroatoms such as nitrogen, sulphur, phosphorous, and oxygen show significant inhibition efficiency. Despite these promising findings about possible corrosion inhibitors, most of these substances are not only expensive but also toxic nonbiodegradable thus causing pollution problems. Hence, these deficiencies have prompted the search for their replacement.

Plants are sources of naturally occurring compounds, some with complex molecular structures and having different chemical, biological, and physical properties. The naturally occurring compounds are mostly used because they are environmentally acceptable, cost effective, and have abundant availability. These advantages are the reason for use of extracts of plants and their products as corrosion inhibitors for metals and alloys under different environment.

Different plant extracts can be used as corrosion inhibitors commonly known as green corrosion inhibitors. Some of them are the following.

Tannins and their derivatives can be used to protect steel, iron, and other tools from corrosion. To protect mild steel in 2 M HCl solutions from corrosion, extracts from leaves can be used. Extracts of tobacco from twigs, stems, and leaves can protect steel and aluminium in saline solutions and strong pickling acids [1, 2]. Extracts from leaves were investigated and found to be effective corrosion inhibitors for mild steel in 2 M HCl solutions. Results for the same are shown in Figure 1.

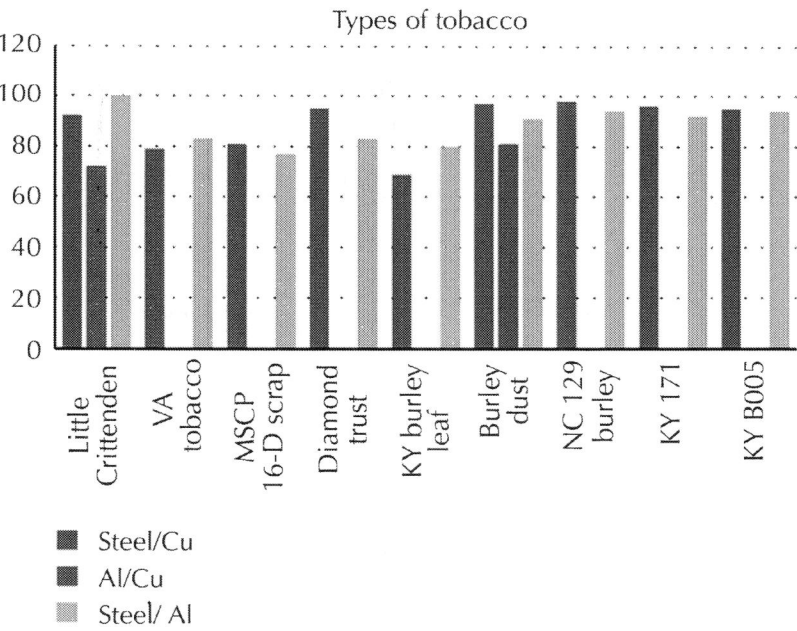

Figure 1: Inhibition efficiency of tobacco extracts for steel/Cu, Al/Cu, and steel/Al galvanic couples in 3.5% NaCl solution as measured by a zero resistance ammeter (ZRA) [1, 2].

It was found that maximum inhibition efficiency is 96% with only 0.01% tobacco concentration (100 ppm). Tobacco extracts contain high concentrations of chemical compounds such as alcohols, polyphenols, nitrogen-containing compounds, terpenes, carboxylic acids, and alkaloids that may exhibit electrochemical activity such as corrosion inhibition [1, 2].

Black pepper, Acacia gum, castor seed, and lignin are also good corrosion inhibitors for steel in acidic media [1, 3]. Mango peel extract is the most effective corrosion inhibitor for Al and Zn, and pomegranate fruit shells extract is most suitable for Cu. It was found that all extracts were more efficiently corrosion inhibitors in HCl solution as compared to H_2SO_4 solution [1, 4]. Aqueous extracts of Eucalyptus leaves protect mild steel and copper in 1 M HCl solution from corrosion [1, 5]. Inhibition efficiency of plant extracts can be tested by various methods such as galvanostatic polarization, mass loss measurements, and surface characterization techniques. SEM studies provide the confirmatory evidence for the protection of mild steel by the inhibitor.

It was found that inhibition efficiency increases with increase in concentration of extract and decreases with increase in temperature.

Extract of leaves of Henna (Lawsonia) acts as a good corrosion inhibitor for carbon steel, nickel, and zinc in acidic, neutral, and alkaline solutions [1, 6]. The degree of inhibition depends on nature of metal and type of medium. For steel and nickel, the inhibition efficiency increased in the order: alkaline < neutral < acidic, while in case of zinc, it increased in the order: acid < alkaline < neutral, thereby reconciling with the observed concept of the Lawsonia extract being a mixed inhibitor [1, 6].

One among the crucial factors for the determination of the inhibition mechanism as well as the performance of the corrosion inhibitor is the solution pH. Most of the inhibitors are pH selective which depends on the molecular structure of the inhibitor, the metal corroding, the active species present in the solution, and the composition of the inhibitor.

Extract of Hibiscus sabdariffa can be used as corrosion inhibitor for mild steel in 2 M HCl and 1 M H_2SO_4 solution [1, 7]. Temperature changes do not affect inhibition performance of Hibiscus sabdariffa in 1 M H_2SO_4 solution.

The application of the acid extract of leaves of Citrus aurantifolia plant on the corrosion inhibition of mild steel in 1 M HCl solution was investigated using weight loss measurement and electrochemical studies [1, 8]. Inhibitive action of the same was tested on adsorption isotherms, and it was found to fit all the models tested, that is, Langmuir, Temkin, Freundlich, Frukin, and Flory - Huggins. This extract also acts as mixed-type inhibitor. A list of various plant materials that have been used as corrosion inhibitors is given in Table 1.

Table 1: Plant materials used as corrosion inhibitors [9]

S. no.	Metal	Medium	Inhibitor	Additive	Method	Findings	Reference
1	Zinc	2 M HCl	Aloe vera	—	Langmuir adsorption isotherm	A first-order kinetics relationship	[10]
2	Mild steel	H2SO4	Aloe vera		Infrared spectrophotometer, thermodynamic adsorption theories and gasometric (hydrogen evolution) methods. The study was conducted at 303 and 333 K	Chemical adsorption isotherm	[11]
3	Concrete steel surface	10 or 23 per cent sodium hydroxide	Banana plant juice taken from paradica and maghraby banana pseudostem		Weight loss method	Anticorrosive materials	[12]
4	Concrete steel surface		Magrabe banana stem		Galvanostatic polarization technique	Mechanical and physic-chemical properties	[13]

5	Mild steel	1 M HCl	Pennyroyal mint		Weight loss measurements, electrochemical polarization, and EIS methods	Cathodic inhibitor, adsorption isotherm	[14]
6	Mild steel	1 M HCl	Justicia gendarussaextract (JGPE)		Weight loss electrochemical techniques. AFM and ESCA	Mixed-type inhibitor. Obeys the Langmuir adsorption isotherm	[15]
7	Mild steel	0.1 M H2SO4	Caffeic acid		Weight loss, potentiodynamic polarisation, electrochemical impedance, and Raman spectroscopy	Controls the anodic reaction	[16]
8	Mild steel	1 M HCl and H2SO4	Combination of leaves and seeds (LVSD) extracts ofphyllanthus amarus		Weight loss and gasometric techniques	Temkin isotherm	[17]
9	Carbon steel	1 M HCl	Aqueous extracts of mango, orange, passion fruit, and cashew peels		Electrochemical impedance, spectroscopy, potentiodynamic polarization curves, weight loss measurements, and surface analysis	Langmuir adsorption isotherm	[18]

10	Carbon steel1	Ethanol	Caffeine		Voltammograms, Tafel plots, and EIS	The standard free energy of adsorption confirms a spontaneous chemical adsorption isotherm step	[19]
11	Al	0.5 M NaOH	Hibiscus sabdariffa leaves		Electrochemical measurements	Mixed-type inhibitor Langmuir and Dubinin Radushkevich isotherm	[20]
12	Al-Zn-Mg alloy	0.5 M NaOH	HibiscusTeterifa		Weight loss measurements	The adsorbed molecules of the alloy, lowers the corrosion rate.	[21]
13	Mild steel	H_2SO_4	Thyme, coriander, hibiscus, anise, black cumin, and garden cress		a.c., d.c., electrochemical techniques, and potentiodynamic polarization	Mixed-type inhibitor	[22]

14	Mild steel		Eucalyptus, hibiscus, and agaricus	Weight loss and polarization methods	Langmuir, Freundlich adsorption isotherm. Agaricus extract was found to be a cathodic inhibitor while extracts of eucalyptus and hibiscus were found to be mixed inhibitors	[23]
15	Mild steel	1 M HCl and 0.5 M H2SO4	Murraya Koenigii leaves	Weight loss, EIS, linear polarization, and potentiodynamic polarization techniques	Langmuir adsorption isotherm (Q, ΔH^*, and DS*)	[24]
16	Mild steel	1 N HCl	Murraya Koenigii	Weight loss, gasometric studies, electrochemical polarization, AC impedance measurements, and SEM studies (30–80°C)	The protective film formed on the surface	[25]
17	Al	2 M HCl	Chromolaena odorata L.	Gasometric and thermometric techniques (30–60°C)	Langmuir adsorption isotherm	[26]

18	Mild steel	H_2SO_4	Ethanol extract of ITHeinsia crinata/IT		Weight loss, thermometric, hydrogen evolution techniques, and IR spectroscopy	Adsorption inhibitor Temkin and Frumkin adsorption	[27]
19	Mild steel	1 M HCl 0.5 H_2SO_4	Dacryodis edulis(DE)		Gravimetric and electrochemical techniques	DE extract was found to inhibit the uniform and localised corrosion of carbon steel in the acidic media	[28]
20	Al	HCl			Weight loss and hydrogen evolution methods	Langmuir adsorption isotherm, activation energies (Ea), activation enthalpy, and activation entropy	[29]
21	Al	0.5 M HCl	Azadirachta indica(AZI) plant	Iodide ions	Potentiodynamic polarization and impedance techniques	Freundlich adsorption isotherm	[30]
22	Mild steel	(60ppm of Cl−)	Aqueous extract of rhizome (Curcuma longaL.) powder	Zn^{2+}	Weight loss method, FTIR, UV fluorescence, and Electrochemical studies	Forms synergistic effect, protective film consists of a Fe^{2+}–curcumin complex and zinc hydroxide (Zn[OH]2)	[31]

23	Al	HCl	Peepal (Ficus religiosa).		Mass loss and thermometric methods	IE dependent upon the concentrations of the inhibitor and the acid	[32]
25	Mild steel	0.1 M HCl	TL and BR inhibitors from green tea and rice bran		Weight loss method, polarization techniques	Cathodic inhibitor	[33]
26	Mild steel	0.2 M HCl	Bark and leaf solution extracts of mango (Mangifera indica)	Ambient temperature	Weight loss method	At 1.0mL/100mL of 0.2 M dilute sulphuric acid concentration gives good IE	[34]
27	Mild steel	HCl	Acid extract ofAndrographis paniculata		Mass loss method, Tafel polarization method, and impedance studies	Plant extract has the potential to serve as corrosion inhibitor	[35]
28	Al NaOH	Abrus precatorius	Ambient temperature		Weight loss and polarization techniques	Suitable adsorption isotherms were tested graphically	[36]
29	Mild steel	H2SO4	Combretum bracteosum		The gravimetric and hydrogen evolution measurements. Temp 30-60°C	Frumkin adsorption isotherm Kinetic parameters calculated, used in chemical cleaning and pickling	[37]

30	Al	1 M HCl	Root of ginseng	Weight loss techniques. Temp 30–60°C	IE 93.1% at 30°C at 50% v/v concentration of ginseng Freundlich adsorption isotherm, thermodynamic parameters calculated	[38]
31	Al	0.5 M NaOH and H2SO4	Vigna unguiculata(VU) extract	Weight loss techniques electrochemical studies. Temp 30 and 60°C	Freundlich and Temkin adsorption isotherms	[39]
32	Mild steel	1 M HCl	Mango, orange, passion fruit, and cashew peels	Electrochemical impedance spectroscopy, potentiodynamic polarization curves, weight loss measurements, and surface analysis	Langmuir adsorption isotherm, IE increases with increasing extract concentration and decreases with temperature	[40]
33	Mild steel	2 M HCl	olive (Olea europaea L.) leaves	Weight loss measurements, Tafel polarization, and cyclic voltammetry	Langmuir adsorption isotherm, olive extract decreases the charge density in the transpassive region	[41]
34	Mild steel	5% HCl	Both aqueous and alcoholic extracts of seven aloe plants	Weight loss measurements	IE 70–82%	[42]

Tannins, organic amino acids, alkaloids, and organic dyes of plant origin have good corrosion-inhibiting abilities. Plant extracts contain many organic compounds, having polar atoms such as O, P, S, and N. These are adsorbed on the metal surface by these polar atoms, and protective films are formed and various adsorption isotherms are obeyed.

The paper incorporates various types of green corrosion inhibitors and their effect on metals. Some important inhibitors in HCl solution, H_2SO_4 solutions, and water solution and effect of temperature and concentration of inhibitors on the process are discussed.

HCL SOLUTION AS MEDIUM

Grape Pomace for Carbon Steel

Acid solutions are widely used in industry, and some of the most important fields of application are acid pickling, chemical cleaning and processing, ore production, and oil well acidification [43–45]. C-steel is one of the most important alloys being used in a wide range of industrial applications. Corrosion problems arise as a result of the interaction between the aqueous solutions and C-steel, especially during the pickling process in which the alloy is brought in contact with highly concentrated acids. This process can lead to economic losses due to the corrosion of the alloy [43,46]. The use of green inhibitors is one of the most practical ways possible for protecting carbon steel from corrosion.

Grape pomace is an industrial waste from wine and juice processing, and it primarily consists of grape seeds, skin and stems (~ 18–20 kg/100 kg of grapes) [43, 47–49]. It was found that grape pomace can effectively protect carbon steel from corrosion in 1 M HCl solution [43].

The inhibition efficiency of C-steel in 1 mol L^{-1} HCl increased with the concentration of crude and concentrated grape pomace extracts and was inversely associated with temperature. Presumably, the inhibitory effect was performed via the adsorption of compounds present in the grape pomace extracts onto the steel surface. Flavonoids are good candidates to explain the corrosion inhibition effects observed for grape pomace extracts. The adsorption of the grape pomace extracts followed a Langmuir adsorption isotherm. The Ea of C-steel dissolution increased in presence of the grape pomace extracts.

SEM revealed the persistence of a smooth surface on C-steel when grape pomace extracts were added, possibly due to the formation of an adsorptive film of phenolic compounds with electrostatic character [43].

Tannin for Mild Steel

Rhizophora racemosa is in abundance in the Mangrove forests of southern Nigeria. The bark of its stem is rich in tannins which can be described as any group of naturally occurring phenolic compounds. Their basic structure consists of garlic acid residues which are linked to glucose via glycosidic bonds [50, 51]. Thus tannins have an array of hydroxyl and carboxyl groups through which the molecules can adsorb on corroding metallic surfaces.

Ferrous materials, especially mild steel, on the other hand are largely used in acidic media in most industries including oil/gas exploration and ancillary activities. During such activities, inhibited hydrochloric acid is widely used in pickling, descaling, and stimulation of oil wells in order to increase oil and gas flow [50]. Tannins from Rhizophora Racemosa was found to be the most effective corrosion inhibitor for mild steel.

Studies on the corrosion behaviour of mild steel electrodes in inhibited hydrochloric acid are described. Conventional weight loss measurements show that a maximum concentration of 140 ppm of tannin from Rhizophora racemosa is required to achieve 72% corrosion inhibition. Similar concentration of tannin : H_3PO_4 in ratio 1 : 1 gave 61% inhibition efficiency, whereas efficiency obtained for phosphoric acid as inhibitor in the same environment was 55%. Corrosion rates obtained over six hours of exposure in 1 M HCl solution at inhibitor concentrations of 140 ppm are 2 mA/cm^2, 2.4 mA/cm^2, 2.6 mA/cm^2, and 6 mA/cm^2 for tannin, tannin/H_3PO_4, and H_3PO_4-inhibited and -uninhibited specimens respectively. Natural atmospheric exposure studies revealed that specimens treated in H_3PO_4 resisted corrosion for three weeks, while tannin-treated specimens suffered corrosion attack after one week of exposure tests [50].

Polyalthia longifolia for Mild Steel

Mild steel finds a lot of application in industries like metal finishing, boiler scale removal, pickling baths, and so forth. It gets rusted when it comes in contact with any acid. Acid solution, mostly HCl, is used to remove any undesirable scale or rust. Corrosion inhibitors are used to prevent the effect of corrosion in such cases. Use of hazardous chemical inhibitors is totally reduced because of environmental regulations. Chromates, phosphates, molybdates, and so forth and a variety of organic compounds containing heteroatoms like nitrogen, sulphur, and oxygen have been investigated as corrosion inhibitors [52–58].

The study shows that acid extract of Polyalthia longifolia (PL) is a good inhibitor for the corrosion of mild steel in HCl. The inhibition efficiency increases with the increase in inhibitor concentration and thus increases the protective action of the inhibitor on mild steel. The compound seems to function as inhibitor by being adsorbed on the metal surface. The inhibitor showed maximum inhibition efficiency of 87.79% at 1.5% v/v inhibitor concentration for an immersion period of 12 hours at 303 K. The % inhibition efficiency increases with increase in temperature, which confirms that PL acts as an effective inhibitor at high temperature also. The adsorption of acid extract of (PL) on the surface of mild steel is spontaneous, endothermic, and consistent with the isotherm models of Langmuir, Temkin, and Freundlich [52].

Flavin Mononucleotide (FMN) for Hot Rolled Steel

Heterocyclic compounds display potential properties for use as corrosion inhibitors due to the presence of nitrogen, oxygen, and sulphur in their ring structure [59–62]. In addition, planarity due to the presence of π electrons and lone pairs of electrons on the heteroatoms contribute to their efficiency as inhibitors.

Flavin mononucleotide (7, 8-dimethyl-10-ribityl-isoalloxazine-5' phosphate monosodium salt dihydrate) is a phosphate monosodium dihydrated salt of Vitamin B2 (Riboflavin). It consists of a heterocyclic isoalloxazine ring attached to the sugar alcohol, ribitol, which is derived from a D(−) pentose sugar (ribose) that contains three antisymmetric carbons and a phosphate monosodium salt [59].

It was found that FMN is a potential inhibitor for corrosion of hot rolled steel in acidic medium. The inhibition efficiency of FMN increases with both concentration and temperature. The inhibitor follows Frumkin isotherm with negative values of ΔG°_{ads}, which signifies that the adsorption is a spontaneous process. High ΔG°_{ads} values indicate that the adsorption takes place by chemisorption at all temperatures except at the lowest temperature, where comprehensive adsorption exists. The Ea values for various concentrations of FMN are lower than Ea for acid, further confirming the role of chemisorption in the adsorption process. Quantum chemical analysis suggests that adsorption of FMN is mainly concentrated around the isoalloxazine ring [59].

WATER SOLUTIONS AS MEDIUM

Gum Exudates from Acacia Species (A. Drepanolobium and A. Senegal) for Mild Steel

Corrosion is a major destructive process affecting the performance of metallic materials in applications in many construction sectors. Corrosion is a naturally occurring phenomenon commonly defined as deterioration of metal surfaces caused by the reaction with the surrounding environmental conditions [63]. The use of the gum exudate from Acacia seyal var seyal as corrosion inhibitor for mild steel in fresh water has been reported [63, 64].

The study shows that gum exudates, from Acacia drepanolobium and Acacia senegal trees, which are natural products, inhibit the corrosion of mild steel in fresh water with A. senegal gum exhibiting better inhibition characteristics compared to Acacia drepanolobium. It was found that the inhibition performances of the Acacia gum exudates are insignificantly affected by temperature rise. Potentiodynamic polarization studies reveal that the gum exudates are mixed-type inhibitors of mild steel corrosion in fresh water with significant reduction of anodic current densities [63].

Asafoetida Extract (ASF) for Mild Steel in Sea Water

Asafoetida is an ingredient of a plant mixture reported to have antidiabetic properties in rats [65,66]. Asafoetida has a broad range of uses in traditional medicine as an antimicrobial, antiepileptic, used for treating chronic bronchitis and whooping cough [65, 67, 68].

It was found that the formulation consisting of 4 mL of ASF and 25 ppm of Zn^{2+} offers 98% inhibition efficiency to carbon steel immersed in sea water. When immersion period increases, corrosion rate also increases. Polarization study reveals that this system formulation acts as a mixed type of inhibitor. The FTIR spectra reveal that the protecting film consists of Fe^{2+} Asafoetida (active ingredient) complex. AFM studies confirm that the surface is smoother. The smoothness of the surface is due to the formation of a compact protective film of Fe^{2+} ASF complex on the metal surface thereby inhibiting the corrosion of carbon steel [65].

Ginger Extract for Steel in Sulfide-polluted Salt Water

Low-grade gram flour, natural honey, onion, potato, gelatin, plant roots, leaves, seeds, and flower gums are some of the good inhibitors. However, most of them have been tested on steel and nickel sheets. Although some studies have been performed on aluminum sheets, the corrosion effect is seen in very mild acidic or basic solutions (mill molar solutions) [69]. It was found that ginger can be effectively used to prevent corrosion of steel in sulphide-polluted salt water. Biological effect of ginger on Escherichia coli was also tested.

Ginger is suggested that it has oxygen donor atoms attached with the proteins and lipids on the bacterial tissues surface making a little activity for it. So it was observed that this inhibitor has no toxicity on the bacterial activity and can be applied on the waste water plants safely without any problems in treating waste water operations [69].

It was found that this extract inhibits the acid-induced corrosion of steel by virtue of adsorption of its components onto the metal surface. The inhibition process is a function of temperature, inhibitor concentration, and the metal as well as inhibitor adsorption abilities which is so much dependent on the number of adsorption sites. The mode of adsorption depends on the type of adsorption (physisorption and chemisorption) observed and could be attributed to the fact that this extract contains many different chemical compounds some of which can adsorb chemically and others adsorb physically. It may be due to the fact that adsorbed organic molecules can influence the behaviour of electrochemical reactions involved in corrosion processes in several ways [69].

Thus it was found that ginger acts as an inhibitor for corrosion of steel in sulfide-polluted salt water. The inhibition efficiency increases with increase in the concentration of the inhibitor. The inhibition is due to the adsorption of the inhibitor molecule on the metal surface by charge transfer or by the diffusion of the inhibitor molecules. The adsorption of these compounds on the metal surface follows Temkin adsorption isotherm. This inhibitor has no biological effect on the activity of Escherichia coli, and can be applied safely on waste water treatment plants.

H_2SO_4 SOLUTION AS MEDIUM

Tannin Extract of Chamaerops Humilis (LF-Ch) Plant for Mild Steel

Tafel polarization curves and electrochemical impedance spectroscopy (EIS) approve that LF-Chextract is an effective corrosion inhibitor for mild steel in 0.5 M sulfuric acid solution +5% EtOH. The inhibition efficiency improved with the increase of LF-Ch extract concentration, whether LF-Ch extract was used alone or in combination with KI. The increase in inhibitor efficiency is generated by the addition of KI to LF-Ch extract. The Tafel polarization curves indicate that both LF-Ch extract is mixed anodic-cathodic type inhibitors. The addition of 0.025% KI to the solution leads to reduction in the essential usage of LF-Ch extract to achieve desirable inhibition efficiency. The values of the inhibition efficiency increased with the immersion time and leads to the formation of a protective film which grows with increasing exposure time [70]. An inhibitor is usually added in small amount in order to slow down the rate of corrosion through the mechanism of adsorption [70–72].

Tryptamine (TA) as a Green Corrosion Inhibitor in Deaerated Sulfuric Acid

Tryptamine (TA), a derivative of the tryptophan, is relatively cheap, nontoxic and easy to produce in purity greater than 99% [73]. TA, a cheap molecule with a very low environmental impact, was found effective in inhibiting ARMCO iron corrosion in deaerated 0.5 M sulphuric acid in the 25–55°C temperature range. Results obtained from potentiodynamic polarisation and electrochemical impedance spectroscopy indicated that TA in the more concentrated solution and at 55°C also chemisorbs. EIS long-time tests (72 h and more) demonstrated that only the 10^{-2} M TA solution attained the maximum protection efficiency both at 25 and 55°C: IP ranged from about 95% to 98% [73].

Essential Oil of Salvia Aucheri Mesatlantica for Steel

Essential oil of aerial parts of Salvia aucheri Boiss. var. mesatlantica was obtained by hydrodistillation and analyzed by GC and GC/MS. The oil was predominated by camphor (49.59%). The inhibitory effect of this essential oil was estimated on the corrosion of steel in 0.5 M H_2SO_4 using electrochemical polarization and weight loss measurements. The corrosion rate of steel is decreased in the presence of natural oil [74]. Chemical analysis shows that camphor can be the major component of S. aucheri mesatlantica oil. Salvia aucheri mesatlantica oil mainly acts as good inhibitor for the corrosion of steel in 0.5 M H_2SO_4. Inhibition efficiency increases with both the concentration of inhibitor and the temperature. The natural oil acts on steel surface as anodic inhibitor. Inhibition efficiency on steel may occur by action of camphor [74].

CONCLUSIONS

Corrosion control of metals is technically, economically, environmentally, and aesthetically important. Corrosion of metals is the major problem in industries. Considering environmental and ecological reasons, green inhibitors are found to be effective. As organic corrosion inhibitors are toxic in nature, so green inhibitors which are biodegradable, without any heavy metals and other toxic compounds, are promoted. Also plant products are inexpensive, renewable, and readily available. The paper discusses some of the important inhibitors in HCl, water, and H_2SO_4 medium and effect of temperature and concentration of inhibitors on the process. Tannins, organic amino acids, alkaloids, and organic dyes of plant origin have good corrosion-inhibiting abilities. Plant extracts contain many organic compounds, having polar atoms such as O, P, S, and N. These are adsorbed on the metal surface by these polar atoms, and protective films are formed, and various adsorption isotherms are obeyed. Corrosion inhibitors can be divided into two broad categories, namely, those that enhance the formation of a protective oxide film through an oxidizing effect and those that inhibit corrosion by selectively adsorbing on the metal surface and creating a barrier that prevents access of corrosive agents to the metal surface. Inhibition efficiency depends on temperature and concentration of inhibitor. Some of the inhibitors are mixed-type inhibitors.

ACKNOWLEDGMENTS

The authors thank the Chemical Engineering Department of Institute of Technology, Nirma University, to provide infrastructure and ample resources needed for the work done.

REFERENCES

1.   J. Buchweishaija, "Phytochemicals as green corrosion inhibitors in various corrosive media a review," Chemistry Department, College of Natural and Applied Sciences, University of Dares Salaam.

2. G. D. Davis, Anthony Von Fraunhofer J, Krebs LA and Dacres CM, 1558, The use of Tobacco extracts as corrosion inhibitors. CORROSION, 2001.

3. K. Srivastava and P. Srivastava, "Studies on plant materials as corrosion inhibitors," British Corrosion Journal, vol. 16, no. 4, pp. 221–223, 1981.

4. R. M. Saleh, A. A. Ismail, and A. A. El Hosary, "Corrosion inhibition by naturally occurring substances. The effect of aqueous extracts of some leaves and fruit peels on the corrosion of steel, aluminum, zinc and copper in acids," British Corrosion Journal, vol. 17, no. 3, pp. 131–135, 1982.

5. K. Pravinar, A. Hussein, G. Varkey, and G. Singh, "Inhibition effect of aqueous extracts ofEucalyptus leaves on the acid corrosion of mild steel and copper," Transaction of the SAEST, vol. 28, no. 1, pp. 8–12, 1993.

6. A. Y. El-Etre, M. Abdallah, and Z. E. El-Tantawy, "Corrosion inhibition of some metals usingLawsonia extract," Corrosion Science, vol. 47, no. 2, pp. 385–395, 2005. ··

7. E. E. Oguzie, "Corrosion inhibitive effect and adsorption behaviour of Hibiscus sabdariffaextract on mild steel in acidic media," Portugaliae Electrochimica Acta, vol. 26, no. 3, pp. 303–314, 2008.

8. R. Saratha, S. V. Priya, and P. Thilagavathy, "Investigation of Citrus aurantiifolia leaves extract as corrosion inhibitor for mild steel in 1 M HCL," E-Journal of Chemistry, vol. 6, no. 3, pp. 785–795, 2009.

9. M. Sangeetha, S. Rajendran, T. S. Muthumegala, and A. Krishnaveni, Green corrosion inhibitors-An Overview.

10. O. K. Abiola and A. O. James, "The effects of Aloe vera extract on corrosion and kinetics of corrosion process of zinc in HCl solution," Corrosion Science, vol. 52, no. 2, pp. 661–664, 2010. ··

11. N. O. Eddy and S. A. Odoemelam, "Inhibition of corrosion of mild steel in acidic medium using ethanol extract of Aloe vera," Pigment and Resin Technology, vol. 38, no. 2, pp. 111–115, 2009. ··

12. M. El-Sayed, O. Y. Mansour, I. Z. Selim, and M. M. Ibrahim, "Identification and utilization of banana plant juice and its pulping liquor as anti-corrosive materials," Journal of Scientific and Industrial Research, vol. 60, no. 9, pp. 738–747, 2001.

13. S. H. Tantawi and I. Z. Selim, "Improvement of concrete properties and reinforcing steel inhibition using a natural product admixture," Journal of Materials Science and Technology, vol. 12, no. 2, pp. 95–99, 1996.

14. A. Bouyanzer, B. Hammouti, and L. Majidi, "Pennyroyal oil from Mentha pulegium as corrosion inhibitor for steel in 1 M HCl," Materials Letters, vol. 60, no. 23, pp. 2840–2843, 2006. · ·

15. A. K. Satapathy, G. Gunasekaran, S. C. Sahoo, K. Amit, and P. V. Rodrigues, "Corrosion inhibition by Justicia gendarussa plant extract in hydrochloric acid solution," Corrosion Science, vol. 51, no. 12, pp. 2848–2856, 2009. · ·

16. F. S. de Souza and A. Spinelli, "Caffeic acid as a green corrosion inhibitor for mild steel,"Corrosion Science, vol. 51, no. 3, pp. 642–649, 2009. · ·

17. P. C. Okafor, M. E. Ikpi, I. E. Uwah, E. E. Ebenso, U. J. Ekpe, and S. A. Umoren, "Inhibitory action of Phyllanthus amarus extracts on the corrosion of mild steel in acidic media,"Corrosion Science, vol. 50, no. 8, pp. 2310–2317, 2008. · ·

18. J. C. da Rocha, J. A. da Cunha Ponciano Gomes, and E. D›Elia, "Corrosion inhibition of carbon steel in hydrochloric acid solution by fruit peel aqueous extracts," Corrosion Science, vol. 52, no. 7, pp. 2341–2348, 2010. · ·

19. L. G. da Trindade and R. S. Gonçalves, "Evidence of caffeine adsorption on a low-carbon steel surface in ethanol," Corrosion Science, vol. 51, no. 8, pp. 1578–1583, 2009. · ·

20. E. A. Noor, "Potential of aqueous extract of Hibiscus sabdariffa leaves for inhibiting the corrosion of aluminum in alkaline solutions," Journal of Applied Electrochemistry, vol. 39, no. 9, pp. 1465–1475, 2009. · ·

21. F. A. Ayeni, V. S. Aigbodion, and S. A. Yaro, "Non-toxic plant extract as corrosion inhibitor for chill cast Al-Zn-Mg alloy in caustic soda solution," Eurasian Chemico-Technological Journal, vol. 9, no. 2, pp. 91–96, 2007.

22. E. Khamis and N. Alandis, "Herbs as new type of green inhibitors for acidic corrosion of steel," Materialwissenschaf Tund Werkstoffiechnik, vol. 33, no. 9, pp. 550–554, 2002.

23. A. Minhaj, P. A. Saini, M. A. Quraishi, and I. H. Farooqi, "A study of natural compounds as corrosion inhibitors for industrial cooling systems," Corrosion Prevention and Control, vol. 46, no. 2, pp. 32–38, 1999.

24. M. A. Quraishi, A. Singh, V. K. Singh, D. K. Yadav, and A. K. Singh, "Green approach to corrosion inhibition of mild steel in hydrochloric acid and sulphuric acid solutions by the extract of Murraya koenigii leaves," Materials Chemistry and Physics, vol. 122, no. 1, pp. 114–122, 2010. · ·

25. A. Sharmila, A. A. Prema, and P. A. Sahayaraj, "Influence of Murraya koenigii (curry leaves) extract on the corrosion inhibition of carbon steel in HCL solution," Rasayan Journal of Chemistry, vol. 3, no. 1, pp. 74–81, 2010.

26. I. B. Obot and N. O. Obi-Egbedi, "An interesting and efficient green corrosion inhibitor for aluminium from extracts of Chlomolaena odorata L. in acidic solution," Journal of Applied Electrochemistry, vol. 40, no. 11, pp. 1977–1984, 2010. · ·

27. N. O. Eddy and A. O. Odiongenyi, "Corrosion inhibition and adsorption properties of ethanol extract of ITHeinsia crinata/IT on mild steel in H_2SO_4," Pigment and Resin Technology, vol. 39, no. 5, pp. 288–295, 2010. · ·

28. E. E. Oguzie, C. K. Enenebeaku, C. O. Akalezi, S. C. Okoro, A. A. Ayuk, and E. N. Ejike, "Adsorption and corrosion-inhibiting effect of Dacryodis edulis extract on low-carbon-steel corrosion in acidic media," Journal of Colloid and Interface Science, vol. 349, no. 1, pp. 283–292, 2010. · ·

29. E. I. Ating, S. A. Umoren, I. I. Udousoro, E. E. Ebenso, and A. P. Udoh, "Leaves extract of ananas sativum as green corrosion inhibitor for aluminium in hydrochloric acid solutions,"Green Chemistry Letters and Reviews, vol. 3, no. 2, pp. 61–68, 2010. · ·

30. S. T. Arab, A. M. Al-Turkustani, and R. H. Al-Dhahiri, "Synergistic effect of Azadirachta Indica extract and iodide ions on the corrosion inhibition of aluminium in acid media,"Journal of the Korean Chemical Society, vol. 52, no. 3, pp. 281–294, 2008.

31.   S. Rajendran, S. Shanmugapriya, T. Rajalakshmi, and A. J. Amal Raj, "Corrosion inhibition by an aqueous extract of rhizome powder," Corrosion, vol. 61, no. 7, pp. 685–692, 2005.

32.   T. Jain, R. Chowdhary, P. Arora, and S. P. Mathur, "Corrosion inhibition of aluminum in hydrochloric acid solutions by peepal (Ficus Religeosa) extracts," Bulletin of Electrochemistry, vol. 21, no. 1, pp. 23–27, 2005.

33.   Z. Liu and G.-L. Xiong, "Preparation and application of plant inhibitors," Corrosion and Protection, vol. 24, no. 4, pp. 146–150, 2003.

34.   C. A. Loto, "The effect of mango bark and leaf extract solution additives on the corrosion inhibition of mild steel in dilute sulphuric acid—part I," Corrosion Prevention and Control, vol. 48, no. 1, pp. 38–41, 2001.

35.   S. P. Ramesh, K. P. Vinod Kumar, and M. G. Sethuraman, "Extract of andrographis paniculata as corrosion inhibitor of mild steel in acid medium," Bulletin of Electrochemistry, vol. 17, no. 3, pp. 141–144, 2001.

36.   R. Rajalakshmi, S. Subhashini, M. Nanthini, and M. Srimathi, "Inhibiting effect of seed extract of Abrus precatorius on corrosion of aluminium in sodium hydroxide," Oriental Journal of Chemistry, vol. 25, no. 2, pp. 313–318, 2009.

37.   P. C. Okafor, I. E. Uwah, O. O. Ekerenam, and U. J. Ekpe, "Combretum bracteosum extracts as eco-friendly corrosion inhibitor for mild steel in acidic medium," Pigment and Resin Technology, vol. 38, no. 4, pp. 236–241, 2009. · ·

38.   I. B. Obot and N. O. Obi-Egbedi, "Ginseng root: a new efficient and effective eco-friendly corrosion inhibitor for aluminium alloy of type AA 1060 in hydrochloric acid solution,"International Journal of Electrochemical Science, vol. 4, no. 9, pp. 1277–1288, 2009.

39.   S. A. Umoren, I. B. Obot, L. E. Akpabio, and S. E. Etuk, "Adsorption and corrosive inhibitive properties of Vigna unguiculata in alkaline and acidic media," Pigment and Resin Technology, vol. 37, no. 2, pp. 98–105, 2008. · ·

40.   J. C. da Rocha, J. A. da Cunha Ponciano Gomes, and E. D'Elia, "Corrosion inhibition of carbon steel in hydrochloric acid solution

by fruit peel aqueous extracts," Corrosion Science, vol. 52, no. 7, pp. 2341–2348, 2010. · ·

41. A. Y. El-Etre, "Inhibition of acid corrosion of carbon steel using aqueous extract of olive leaves," Journal of Colloid and Interface Science, vol. 314, no. 2, pp. 578–583, 2007. · ·

42. R. M. Saleh, M. A. Abd El Alim, and A. A. El Hosary, "Corrosion inhibition by naturally occurring substances: constitution and inhibiting property of Aloe plants," Corrosion Prevention and Control, vol. 30, no. 1, pp. 9–10, 1983.

43. J. C. Da Rocha, G. J. A. Ponciano C, E. D. Elia et al., "Grape pomace extracts as green corrosion inhibitors for carbon steel in hydrochloric acid solutions," International Journal of Electrochemical Science, vol. 7, pp. 11941–11956.

44. X.-H. Li, S.-D. Deng, and H. Fu, "Inhibition by Jasminum nudiflorum Lindl. leaves extract of the corrosion of cold rolled steel in hydrochloric acid solution," Journal of Applied Electrochemistry, vol. 40, no. 9, pp. 1641–1649, 2010. ·

45. A. U. Ezeoke, O. G. Adeyemi, O. A. Akerele, and N. O. Obi-Egbedi, "Computational and experimental studies of 4-aminoantipyrine as corrosion inhibitor for mild steel in sulphuric acid solution," International Journal of Electrochemical Science, vol. 7, no. 1, pp. 534–553, 2012.

46. A. Y. El-Etre, "Inhibition of C-steel corrosion in acidic solution using the aqueous extract of zallouh root," Materials Chemistry and Physics, vol. 108, no. 2-3, pp. 278–282, 2008. ·

47. L. M. A. S. de Campos, F. V. Leimann, R. C. Pedrosa, and S. R. S. Ferreira, "Free radical scavenging of grape pomace extracts from Cabernet sauvingnon (Vitis vinifera)," Bioresource Technology, vol. 99, no. 17, pp. 8413–8420, 2008. · ·

48. M. Spanghero, A. Z. M. Salem, and P. H. Robinson, "Chemical composition, including secondary metabolites, and rumen fermentability of seeds and pulp of Californian (USA) and Italian grape pomaces," Animal Feed Science and Technology, vol. 152, no. 3-4, pp. 243–255, 2009. · ·

49. M. A. Bustamante, R. Moral, C. Paredes, A. Pérez-Espinosa, J. Moreno-Caselles, and M. D. Pérez-Murcia, Waste Management, 2008.

50. M. Oki, E. Charles, C. Alaka, and T. K. Oki, Corrosion Inhibition of Mild Steel in Hydrochloric Acid By Tannins From Rhizophora Racemosa Materials Sciences and Applications, vol. 2, 2011.

51. G. I. Nonaka, "The isolation and structure elucidation of tannins," Pure and Applied Chemistry, vol. 6, no. 3, pp. 357–360, 1989.

52. V. G. Vasudha and K. Shanmuga Priya, "Polyalthia longifolia as a corrosion inhibitor for mild steel in HCl solution," Research Journal of Chemical Sciences, vol. 3, no. 1, pp. 21–26, 2013.

53. S. A. M. Refaey, "Inhibition of steel pitting corrosion in HCl by some inorganic anions,"Applied Surface Science, vol. 240, no. 1–4, pp. 396–404, 2005. · ·

54. M. A. Quraishi and H. K. Sharma, "Thiazoles as corrosion inhibitors for mild steel in formic and acetic acid solutions," Journal of Applied Electrochemistry, vol. 35, no. 1, pp. 33–39, 2005. · ·

55. H. Ashassi-Sorkhabi, B. Shaabani, and D. Seifzadeh, "Corrosion inhibition of mild steel by some schiff base compounds in hydrochloric acid," Applied Surface Science, vol. 239, no. 2, pp. 154–164, 2005. · ·

56. M. Bouklah, A. Ouassini, B. Hammouti, and A. El Idrissi, "Corrosion inhibition of steel in sulphuric acid by pyrrolidine derivatives," Applied Surface Science, vol. 252, no. 6, pp. 2178–2185, 2006. · ·

57. E. E. Oguzie, B. N. Okolue, E. E. Ebenso, G. N. Onuoha, and A. I. Onuchukwu, "Evaluation of the inhibitory effect of methylene blue dye on the corrosion of aluminium in hydrochloric acid," Materials Chemistry and Physics, vol. 87, no. 2-3, pp. 394–401, 2004. · ·

58. S. A. Ali, M. T. Saeed, and S. U. Rahman, "The isoxazolidines: a new class of corrosion inhibitors of mild steel in acidic medium," Corrosion Science, vol. 45, no. 2, pp. 253–266, 2003. · ·

59. S. M. Bhola, G. Singh, and B. Mishra, "Flavin mononucleotide as a corrosion inhibitor for hot rolled steel in hydrochloric acid," International Journal of Electrochemical Science, vol. 8, pp. 5635–5642, 2013.

60. M. A. Quraishi and R. Sardar, "Dithiazolidines—a new class of heterocyclic inhibitors for prevention of mild steel corrosion in

hydrochloric acid solution," Corrosion, vol. 58, no. 2, pp. 103–107, 2002.

61. S. L. Granese, B. M. Rosales, C. Oviedo, and J. O. Zerbino, "The inhibition action of heterocyclic nitrogen organic compounds on Fe and steel in HCl media," Corrosion Science, vol. 33, no. 9, pp. 1439–1453, 1992.

62. S. N. Banerjee and S. Misra, "1,10,-phenanthroline as corrosion inhibitor for mild steel in sulfuric acid solution," Corrosion, vol. 45, no. 9, pp. 780–783, 1989.

63. J. Buchweishaija, Plants As a Source of Green Corrosion Inhibitors: The Case of Gum Exudates From Acacia Species, Chemistry Department, College of Natural and Applied Science.

64. J. Buchweishaija and G. S. Mhinzi, "Natural products as a source of environmentally friendly corrosion inhibitors: the case of gum exudate from Acacia seyal var. seyal," Portugaliae Electrochimica Acta, vol. 26, no. 3, pp. 257–265, 2008.

65. M. Sangeetha, S. Rajendran, J. Sathiyabama, and P. Prabhakar, "Asafoetida extract (ASF) as green corrosion inhibitor for mild steel in sea water," International Research Journal of Environment Sciences, vol. 1, no. 5, pp. 14–21, 2012.

66. F. M. Al-Awadi, M. A. Khattar, and K. A. Gumaa, "On the mechanism of the hypoglycaemic effect of a plant extract," Diabetologia, vol. 28, no. 7, pp. 432–434, 1985.

67. K. Srinivasan, "Role of spices beyond food flavoring: nutraceuticals with multiple health effects," Food Reviews International, vol. 21, no. 2, pp. 167–188, 2005.

68. M. Z. Abdin and Y. P. Abdin, Abrol, ISBN 81-7319-707-5, Published Alpha Science IntʹL Ltd. TraditionaL Systems of Medicine, 2005.

69. A. E. -A. S. Fouda, A. A. Nazeer, M. Ibrahim, and M. Fakih, "Ginger extract as green corrosion inhibitor for steel in sulfide polluted salt water," Journal of the Korean Chemical Society, vol. 57, no. 2, pp. 272–278, 2013.

70. O. Benali, H. Benmehdi, O. Hasnaoui, C. Selles, and R. Salghi, "Green corrosion inhibitor: inhibitive action of tannin extract of Chamaerops humilis plant for the corrosion of mild steel in 0. 5M H_2SO_4," Journal of Materials and Environmental Science, vol. 4, no. 1, pp. 127–138, 2013.

71. N. O. Eddy, "Inhibitive and adsorption properties of ethanol extract of Colocasia esculentaleaves for the corrosion of mild steel in H_2SO_4," International Journal of Physical Sciences, vol. 4, no. 4, pp. 165–171, 2009.

72. A. Bouyanzer and B. Hammouti, "A study of anti-corrosive effects of Artemisia oil on steel,"Pigment and Resin Technology, vol. 33, no. 5, pp. 287–292, 2004. · ·

73. G. Moretti, F. Guidi, and G. Grion, "Tryptamine as a green iron corrosion inhibitor in 0.5 M deaerated sulphuric acid," Corrosion Science, vol. 46, no. 2, pp. 387–403, 2004.

74. M. Znini, L. Majidi, A. Bouyanzer et al., "Essential oil of Salvia aucheri mesatlantica as a green inhibitor for the corrosion of steel in 0.5 M H_2SO_4," Arabian Journal of Chemistry, vol. 5, no. 4, pp. 467–474, 2010. · ·

Green Inhibitors for Corrosion Protection of Metals and Alloys: An Overview

B. E. Amitha Rani and Bharathi Bai J. Basu

Surface Engineering Division, CSIR-National Aerospace Laboratories, Bangalore 560037, India

ABSTRACT

Corrosion control of metals is of technical, economical, environmental, and aesthetical importance. The use of inhibitors is one of the best options of protecting metals and alloys against corrosion. The environmental toxicity of organic corrosion inhibitors has prompted the search for green corrosion inhibitors as they are biodegradable, do not contain heavy metals or other toxic compounds. As in addition to being environmentally friendly and ecologically acceptable, plant products are inexpensive, readily available and renewable. Investigations of corrosion inhibiting abilities of tannins, alkaloids, organic,amino acids, and organic dyes of plant origin are of interest. In recent years, sol-gel coatings doped with inhibitors show real promise. Although substantial research has been devoted to corrosion inhibition by plant extracts, reports on the detailed mechanisms of the adsorption process and identification of the active ingredient are still scarce. Development of computational modeling backed by wet experimental results would help to fill this void and help understand the mechanism of inhibitor action, their adsorption patterns, the inhibitor-metal surface interface and aid the development of designer inhibitors with an understanding of the time required for the release of self-healing inhibitors. The present paper consciously restricts itself mainly to plant materials as green corrosion inhibitors.

INTRODUCTION

Corrosion is the deterioration of metal by chemical attack or reaction with its environment. It is a constant and continuous problem, often difficult to eliminate completely. Prevention would be more practical and achievable than complete elimination. Corrosion processes develop fast after disruption of the protective barrier and are accompanied by a number of reactions that change the composition and properties of both the metal surface and the local environment, for example, formation of oxides, diffusion of metal cations into the coating matrix, local pH changes, and electrochemical potential. The study of corrosion of mild steel and iron is a matter of tremendous theoretical and practical concern and as such has received a considerable amount of interest. Acid solutions, widely used in industrial acid cleaning, acid descaling, acid pickling, and oil well acidizing, require the use of corrosion inhibitors in order to restrain their corrosion attack on metallic materials.

CORROSION INHIBITORS

Over the years, considerable efforts have been deployed to find suitable corrosion inhibitors of organic origin in various corrosive media [1–4]. In acid media, nitrogen-base materials and their derivatives, sulphur-containing compounds, aldehydes, thioaldehydes, acetylenic compounds, and various alkaloids, for example, papaverine, strychnine, quinine, and nicotine are used as inhibitors. In neutral media, benzoate, nitrite, chromate, and phosphate act as good inhibitors. Inhibitors decrease or prevent the reaction of the metal with the media. They reduce the corrosion rate by

- adsorption of ions/molecules onto metal surface,
- increasing or decreasing the anodic and/or cathodic reaction,
- decreasing the diffusion rate for reactants to the surface of the metal,
- decreasing the electrical resistance of the metal surface.
- inhibitors that are often easy to apply and have in situ application advantage.

Several factors including cost and amount, easy availability and most important safety to environment and its species need to be considered when choosing an inhibitor.

Organic Inhibitors

Organic inhibitors generally have heteroatoms. O, N, and S are found to have higher basicity and electron density and thus act as corrosion inhibitor. O, N, and S are the active centers for the process of adsorption on the metal surface. The inhibition efficiency should follow the sequence $O < N < S < P$. The use of organic compounds containing oxygen, sulphur, and especially nitrogen to reduce corrosion attack on steel has been studied in some detail. The existing data show that most organic inhibitors adsorbed on the metal surface by displacing water molecules on the surface and forming a compact barrier. Availability of nonbonded (lone pair) and p-electrons in inhibitor molecules facilitate electron transfer from the inhibitor to the metal. A coordinate covalent bond involving transfer of electrons from inhibitor to the metal surface may be formed. The strength of the chemisorption bond depends upon the electron density on the donor atom of the functional group and also the polarizability of the group. When an H atom attached to the C in the ring is replaced by a substituent group ($-NH_2$, $-NO_2$, $-CHO$, or $-COOH$) it improves inhibition [4]. The electron density in the metal at the point of attachment changes resulting in the retardation of the cathodic or anodic reactions. Electrons are consumed at the cathode and are furnished at the anode. Thus, corrosion is retarded. Straight chain amines containing between three and fourteen carbons have been examined. Inhibition increases with carbon number in the chain to about 10 carbons, but, with higher members, little increase or decrease in the ability to inhibit corrosion occurs. This is attributed to the decreasing solubility in aqueous solution with increasing length of the hydrocarbon chain. However, the presence of a hydrophilic functional group in the molecule would increase the solubility of the inhibitors.

The performance of an organic inhibitor is related to the chemical structure and physicochemical properties of the compound like functional groups, electron density at the donor atom, p-orbital character, and the electronic structure of the molecule. The inhibition could be due to (i) Adsorption of the molecules or its ions on anodic and/or cathodic sites, (ii) increase in cathodic and/or anodic over voltage, and (iii) the formation of a protective barrier film. Some factors that contribute to the action of inhibitors are

- chain length,
- size of the molecule,
- bonding, aromatic/conjugate,
- strength of bonding to the substrate,
- cross-linking ability,
- solubility in the environment.

The role of inhibitors is to form a barrier of one or several molecular layers against acid attack. This protective action is often associated with chemical and/or physical adsorption involving a variation in the charge of the adsorbed substance and transfer of charge from one phase to the other. Sulphur and/or nitrogen-containing heterocyclic compounds with various substituents are considered to be effective corrosion inhibitors. Thiophene, hydrazine derivatives offer special affinity to inhibit corrosion of metals in acid solutions. Inorganic substances such as phosphates, chromates, dichromates, silicates, borates, tungstates, molybdates, and arsenates have been found effective as inhibitors of metal corrosion. Pyrrole and derivatives are believed to exhibit good protection against corrosion in acidic media. These inhibitors have also found useful application in the formulation of primers and anticorrosive coatings, but a major disadvantage is their toxicity and as such their use has come under severe criticism. Among the alternative corrosion inhibitors, organic substances containing polar functions with nitrogen, sulphur, and/or oxygen in the conjugated system have been reported to exhibit good inhibiting properties. The inhibitive characteristics of such compounds derive from the adsorption ability of their molecules, with the polar group acting as the reaction center for the adsorption process. The resulting adsorbed film acts as a barrier that separates the metal from the corrodent, and efficiency of inhibition depends on the mechanical, structural, and chemical characteristics of the adsorption layers formed under particular conditions.

Inhibitors are often added in industrial processes to secure metal dissolution from acid solutions. Standard anti corrosion coatings developed till date passively prevent the interaction of corrosion species and the metal. The known hazardous effects of most synthetic organic inhibitors and the need to develop cheap, nontoxic and ecofriendly processes have now urged researchers to focus on the use of natural products. Increasingly, there is a need to develop sophisticated new generation coatings for improved performance, especially in view of Cr VI being banned and labeled as a carcinogen. The use of inhibitors is one of the best options of protecting metals against corrosion. Several inhibitors in use are either synthesized from cheap raw material or chosen from compounds having heteroatoms in their aromatic or long-chain carbon system. However, most of these inhibitors are toxic to the environment. This has prompted the search for green corrosion inhibitors.

GREEN INHIBITORS

Green corrosion inhibitors are biodegradable and do not contain heavy metals or other toxic compounds. Some research groups have reported the successful use of naturally occurring substances to inhibit the corrosion of metals in acidic and alkaline environment. Delonix regia extracts inhibited the corrosion of aluminum in hydrochloric acid solutions [5], rosemary leaves were studied as corrosion inhibitor for the Al + 2.5Mg alloy in a 3% NaCl solution at 25°C [6], and El-Etre investigated natural honey as a corrosion inhibitor for copper [7] and investigated opuntia extract on aluminum [8]. The inhibitive effect of the extract of khillah (Ammi visnaga) seeds on the corrosion of SX 316 steel in HCl solution was determined using weight loss measurements as well as potentiostatic technique. The mechanism of action is attributed to the formation of insoluble complexes as a result of interaction between iron cations, and khellin [9] and Ebenso et al. showed the inhibition of corrosion with ethanolic extract of African bush pepper (Piper guinensis) on mild steel [10];Carica papaya leaves extract [11]; neem leaves extract (Azadirachta indica) on mild steel in H_2SO_4 [12]. Zucchi and Omar investigated plant extracts of Papaia, Poinciana pulcherrima, Cassia occidentalis, and Datura stramonium seeds and Papaya, Calotropis procera B, Azadirachta indica, and Auforpio turkiale sap for their corrosion inhibition potential and found that all extracts except those of Auforpio turkiale and Azadirachta indica reduced the corrosion of steel with an efficiency of 88%–96% in 1 N HCl and with a slightly lower efficiency in 2 N HCl. They attributed the effect to the products of the hydrolysis of the protein content of these plants [13]; Umoren et al. [14], studied the corrosion inhibition of mild steel in H_2SO_4 in the presence of gum arabic (GA) (naturally occurring polymer) and polyethylene glycol (PEG) (synthetic polymer). It was found that PEG was more effective than gum arabic.

Yee [15] determined the inhibitive effects of organic compounds, namely, honey and Rosmarinus officinalis L on four different metals—aluminium, copper, iron, and zinc, each polarized in two different solutions, that is, sodium chloride and sodium sulphate. The experimental approach employed potentiodynamic polarization method. The best inhibitive effect was obtained when zinc was polarised in both honey-added sodium chloride and sodium sulphate solutions. Rosemary extracts showed some cathodic inhibition when the metal was polarized in sodium chloride solution. This organic compound, however, displayed less anodic inhibition when compared with honey. The main chemical components of rosemary include borneol, bornyl acetate, camphor, cineole, camphene, and alpha-pinene. Chalchat et al. [16], reported that oils of rosemary were found to be rich in 1,8-cineole, camphor, bornyl acetate, and high amount of hydrocarbons. Recently, work has been emphasized on the use of Rosmarinus officinalis L as corrosion inhibitor for Al-Mg corrosion in chloride solution [6]. It is believed that the catechin fraction present in the rosemary extracts contributes to the inhibitive properties that act upon the alloy. Ouariachi et al. [17] also reported the inhibitory action ofRosmarinus officinalis oil as green corrosion inhibitors on C38 steel in 0.5 M H_2SO_4.

Odiongenyi et al. [18] reported that the ethanolic extract of Vernonia amygdalina appears to be a good inhibitor for the corrosion of mild steel in H_2SO_4 and action is by classical Langmuir adsorption isotherm.

The effect of addition of halides (KCl, KBr, and KI) was also studied, and the results obtained indicated that the increase in efficiency was due to synergism [13]. Umoren et al. also investigated the corrosion properties of Raphia hookeri exudates gum—halide mixtures for aluminum corrosion in acidic medium [16]. Raphia hookeri exudates gum obeys Freundlich, Langmuir, and Temkin adsorption isotherms. Phenomenon of physical adsorption is proposed. Abdallah also tested the effect of guar gum on carbon steel. It is proposed that it acts as a mixed type inhibitor [14]. The mechanism of action of C-steel by guar gum is due to the adsorption at the electrode/solution interface. Guar gum is a polysaccharide compound containing repeated heterocyclic pyrane moiety as shown in Scheme 1. The presence of heterooxygen atom in the structure makes possible its adsorption by coordinate type linkage through the transfer of lone pairs of electron of oxygen atoms to the steel surface, giving a stable chelate five-membered ring with ferrous ions. The chelation between O1 and O2 with Fe++ seems to be impossible due to proximity factor presented as in Scheme 1:

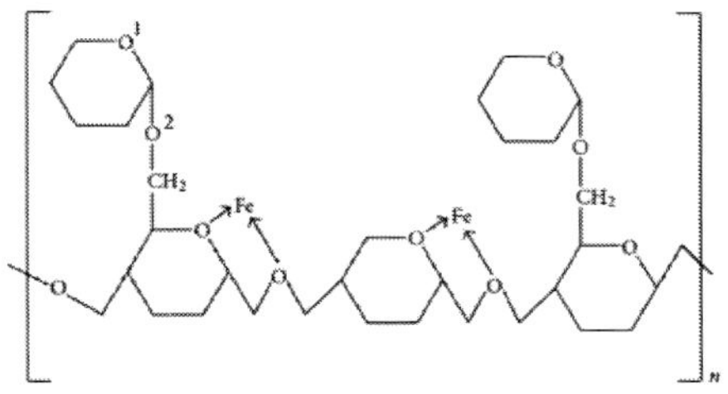

Scheme 1: Guar gum.

The simultaneous adsorption of oxygen atoms forces the guar gum molecule to be horizontally oriented at the metal surface, which led to increasing the surface coverage and consequently protection efficiency even in the case of low inhibitor concentrations.

Okafor et al. looked into the extracts of onion (Allium sativum), Carica papaya extracts, Garcinia kola, and Phyllanthus amarus [19–22]. El-Etre, Abdallah M used Natural honey as corrosion inhibitor for metals and alloys. II C-steel in high saline water [23]. Jojoba oil has also been evaluated [24]. Artemisia oil has been investigated for it is anticorrosion properties [25]. Oguzie and coworkers evaluated Telfaria occidentalis,Occinum viridis, Azadirachta indica, and Sanseviera trifasciata extracts [26–29]. Benda-hou et al., studied using the extracts of rosemary in steel [27], and Sethuraman studied Datura [30]. Recently, studies on the use of some drugs as corrosion inhibitors have been reported by some researchers [31, 32]. Most of these drugs are heterocyclic compounds and were found to be environmentally friendly, hence, they have great potentials of competing with plant extracts. According to Eddy et al. drugs are environmentally friendly because they do not contain heavy metals or other toxic compound. In view of this adsorption and inhibitive efficiencies of ACPDQC (5-amino-1-cyclopropyl-7-[(3R, 5S) 3, 5-dimethylpiperazin-1-YL]-6,8-difluoro-4-oxo-uinoline-3-carboxylic acid), on mild steel corrosion have been studied and found to be effective.

Eddy et al. [33] studied inhibition of the corrosion of mild steel by ethanol extract of Musa species peel using hydrogen evolution and thermometric methods of monitoring corrosion. Inhibition efficiency of the extract was found to vary with concentration, temperature, period of immersion, pH, and electrode potentials. Adsorption of Musa species extract on mild steel surface was spontaneous and occurred according to Langmuir and Frumkin adsorption isotherms and also physical adsorption. Deepa Rani and Selvaraj [34] report the inhibition efficacy of Punica granatum extract on the corrosion of Brass in 1 N HCl evaluated by mass loss measurements at various time and temperature. Langmuir and Frumkin adsorption isotherms appear to be the mechanism of adsorption based on the values of activation energy, free energy of adsorption. Few researchers have summarized the effect of plant extracts on corrosion [35–38].

Efforts to find naturally organic substances or biodegradable organic materials to be used as corrosion inhibitors over the years have been intensified. Several reports are available on the various natural products used as green inhibitors as shown in Tables 1 and 2. Low-grade gram flour, natural honey, onion, potato, gelatin, plant roots, leaves, seeds, and flowers gums have been reported as good inhibitors. However, most of them have been tested on steel and nickel sheets. Although some studies have been performed on aluminum sheets, the corrosion effect is seen in very mild acidic or basic solutions (millimolar solutions).

Table 1: Green inhibitors used for corrosion inhibition of steel

Sl. no.	Metal	Inhibitor source	Active ingredient	References
(1)	Steel	Tamarind		[39]
(2)	Steel	Tea leaves		[40]
(3)	Steel	Pomegranate juice and peels		[41]
(4)	Steel	Emblica officinalis		[42]
(5)	Steel	Terminalia bellerica		[43]
(6)	Steel	Eucalyptus oil	Monomtrene 1,8-cineole	[44]
(7)		Rosemary		[45]
(8)	C-steel, Ni, Zn	Lawsonia extract (Henna)	Lawsone (2-hydroxy-1, 4-napthoquinone resin and tannin, coumarine, Gallic, acid, and sterols)	[46]
(9)	Mild steel	Gum exudate	Hexuronic acid, neutral sugar residues, volatile monoterpenes, canaric and related triterpene acids, reducing and nonreducing sugars	[47]
(10)	Mild steel	Musa sapientum peels (Banana peels)		[48]
(11)	Carbon steel	Natural amino acids—alanine, glycine, and leucine		[48]
(12)	Steel	Natural amino acids		[15]
(13)	Mild steel	Garcinia kola seed	Primary and secondary amines Unsaturated fatty acids and biflavnone	[49]
(14)	Steel	Auforpio turkiale	Protein hydrolysis	[50]
(15)	Steel	Azydracta indica	Protein hydrolysis	[51]
(16)	Steel	Aloe leaves		[52]

(17)	Steel	Mango/orange peels		[53]
(18)	Steel	Hibiscus sabdariffa (Calyx extract) in 1 M H_2SO_4 and 2 M HCl solutions, Stock 10–50%	Molecular protonated organic species in the extract. Ascorbic acid, amino acids, flavonoids, Pigments and carotene	[54]

Table 2: Green Inhibitors used for corrosion inhibition of aluminum, aluminum alloys, and other metals and alloys

Sl. no.	Metal	Inhibitor source	Active ingredient	References
(1)	Al	CeCl$_3$ and mercaptobenzothiazole (MBT)		[55]
(2)	Al, steel	Aqueous extract of tobacco plant and its parts	Nicotine	[56]
(3)	Al	Vanillin		[57]
(4)	Al-Mg alloy	Aqueous extract of Rosmarinus officinalis—Neutral phenol subfraction of the aqueous extract	Catechin	[58]
(5)	Al	Sulphates/molybdates and dichromates as passivators		[59]
(6)	Al	Amino and polyamino acids—aspartic acid		[6]
(7)	Al	Pyridine and its selected derivatives (symmetric collidine and 2,5-dibrompyridine)		[60]
(8)	Al	Citric acid		[61]
(9)	Fe, Al	Benzoic acid		[62]
(10)	Al	Rutin and quercetin		[63]
(11)	Al			US Patent 5951747
(12)	Al	Polybutadieonic acid		[64]

(13)	Al and Zn	Saccharides—mannose and fructose		[65]
(14)	Al, Al-6061 and Al-Cu	Neutral solutions using sulphates, molybdates, and dichromates		[66]
(15)	Al	Vernonia amygdalina (Bitter leaf)		[67]
(16)	Al	Prosopis—cineraria (khejari)		[60]
(17)	Al	Tannin beetroot		[68]
(18)	Al	Saponin		[69]
(19)	Al	Acacia concianna		[70]
(20)	Al and Zn	Saccharides		[71]
(21)	Al	Opuntia (modified stems cladodes)	Polysaccharide (mucilage and pectin)	[72]
(22)	Al-Mg alloy	Rosmarinus officinalis		[8]
(23)	Zn	Metal chelates of citric acid		[61]
(24)	Zn	Onion juice	S-containing acids (glutamyl peptides) S-(1-propenyl) L-cysteine sulfoxide, and S-2-carboxypropyl glutathione	[63]
(25)	Sn	Natural honey (acacia chestnut)		[64]
(26)	Sn	Black radish	120	[8]

Mechanism of Action of Green Inhibitors

Many theories to substantiate the mode of action of these green inhibitors have been put forth by several workers. Mann has suggested that organic substances, which form onium ions in acidic solutions, are adsorbed on the cathodic sites of the metal surface and interfere with the cathodic reaction.

Various mechanisms of action have been postulated for the corrosion inhibition property of the natural products.

Argemone Mexicana

It is a contaminant of mustard seeds contain an alkaloid berberine which has a long-chain of aromatic rings, an N atom in the ring, and, at several places H atoms attached to C are replaced by groups, –CH, –OCH$_3$, and –O. The free electrons on the O and N atoms form bonds with the electrons on the metal surface. Berberine in water ionizes to release a proton, thus the now negatively charged O atom helps to free an electron on the N atom and forms a stronger bond with the metallic electrons. These properties confer good inhibition properties to Argemone mexicana (Scheme 2).

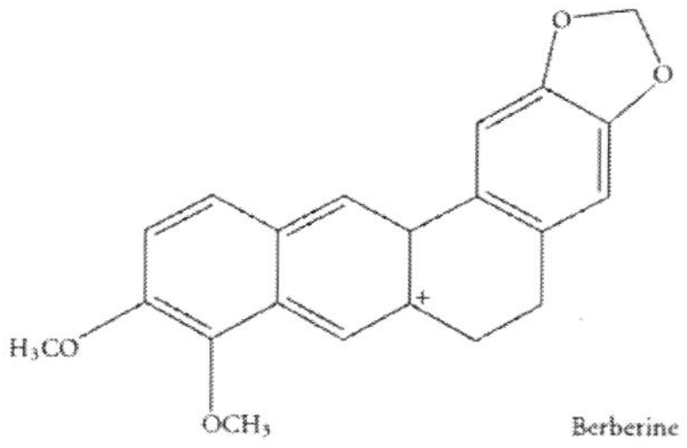

H$_3$CO

OCH$_3$ Berberine

Scheme 2: Berberine.

Garlic

It contains allyl propyl disulphide. Probably, this S-containing unsaturated compounds affects the potential cathodic process of steel.

Carrot

It contains pyrrolidine in aqueous media, pyrrolidine ionizes, and the N atom acquires a negative charge, and the free electrons on N possess still higher charge, resulting in stronger bond formation at N Carrot does not ionize in acidic media and thus does not protect in acids (Scheme 3).

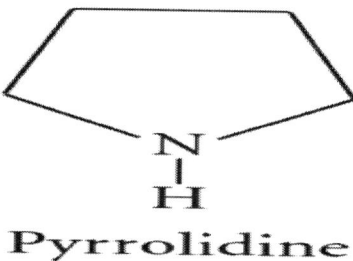

Pyrrolidine

Scheme 3: Pyrrolidine.

Castor Seed

They contain the alkaloid ricinine. The N atom is in the ring attachment of the –OCH$_3$ (Scheme 4).

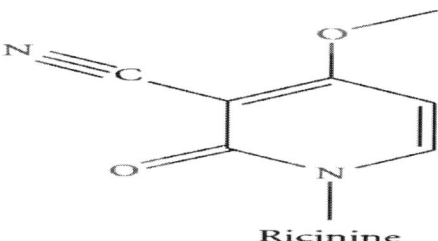

Ricinine

Scheme 4: Ricinine.

Black Pepper

Quraishi et al. [73] studied corrosion inhibition of mild steel in hydrochloric solution by black pepper extract (Piper nigrum family: Piperaceae) by mass loss measurements, potentiodynamic polarisation, and electrochemical impedance spectroscopy (EIS). Black pepper extract gave maximum inhibition efficiency (98%) at 120 ppm at 35°C for mild steel in hydrochloric acid medium. Electrochemical evaluation revealed it to be a mixed-type inhibitor and that charge transfer controls the corrosion process. The corrosion inhibition property was attributed to an alkaloid "Piperine".

Fennel Seeds

Essential oil from fennel (Foeniculum vulgare) (FM) was tested as corrosion inhibitor of carbon steel in 1 M HCl using electrochemical impedance spectroscopy (EIS), Tafel polarisation methods, and weight loss measurements [74]. The increase of the charge-transfer resistance (R_{ct}) with the oil concentration supports the molecules of oil adsorption on the metallic surface. The polarization plots reveal that the addition of natural oil shifts the cathodic and anodic branches towards lower currents, indicative of a mixed-type inhibitor. The analysis of FM oil, obtained by hydrodistillation, using Gas Chromatography (GC) and Gas Chromatography/Mass Spectrometry (GC/MS) showed that the major components were limonene (20.8%) and pinene (17.8%). Interestingly, the composition of FM oil was variable according to the area of harvest and the stage of development. The analysis allowed the identification of 21 components which accounted for 96.6% of the total weight. The main constituents were limonene (20.8%) and pinene (17.8%) followed by myrcene (15%) and fenchone (12.5%). The adsorption of these molecules could take place via interaction with the vacant d-orbitals of iron atoms (chemisorption). It is logical to assume that such adsorption is mainly responsible for the good protective properties by a synergistic effect of various molecules [74–76].

Garcinia Mangostana

Vinod Kumar et al. [77] studied the corrosion inhibition of acid extract of the pericarp of the fruit of G. mangostana on mild steel in hydrochloric acid medium. G. mangostana, colloquially known as "the mangosteen", is a tropical evergreen tree. Mangosteen fruit, (Figure 1) on ripening the fruit, turns from green to purple in colour.

(a)

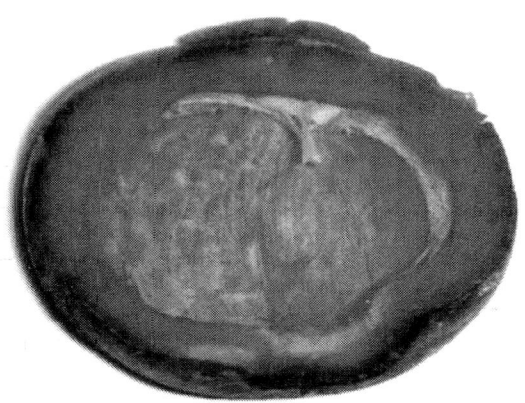

(b)

Figure 1: (a) Mangostana fruit. (b) Pericarp.

The extract of the pericarp of G. mangostana contains oxygenated prenylated xanthones, 8-hydroxycudraxanthone G and mangostingone [7-methoxy-2-(3-   methyl-2-butenyl)-8-(3-methyl-2-oxo-3-butenyl)-1,3,6-trihydroxyxanthone, along with other xanthones such as cudraxanthone G, 8-deoxygartanin, garcimangosone B, garcinone D, garcinone E, gartanin, 1-isomangostin, á-mangostin, γ-mangostin, mangostinone, smeathxanthone A, and tovophyllin A [77, 78]. Electrochemical parameters such as Ecorr, $_{a'}$ and β_c indicate the mixed mode of inhibition, but predominantly cathodic. IR analysis and impedance studies indicate that the adsorption on the metal surface is due to the heteroatoms present in the organic constituents of the extract of G. mangostana.

Ipomea Involcrata

Obot et al.[79] studied the corrosion inhibition efficiency of Ipomoea involcrata (IP) (family: Convolulaceae) leaf extract on aluminium. It is a common ornamental vine with heart-shaped and bright white pink or purple flowers that has a long history of use in central to southern Mexico. The plant has been shown to contain mainly d-lysergic acid amide (LSA) (Figure 2) and small amounts of other alkaloids, namely, chanoclavine, elymoclavine, and ergometrine, and d-isolysergic acid amide [79]. D-lysergic acid amide (LSA) (Figure 2) contains N and O in their structure including π-electrons which are required for corrosion inhibiting effects. Probably, chanoclavine, elymoclavine, ergometrine, d-isolysergic acid amide, and other ingredients of the plant extracts synergistically increase the strength of the layer formed by the d-lysergic acid amide (LSA). Thus, the formation of a strong physisorbed layer between the metal surface and the phytoconstituents of the plant extract could be the cause of the inhibitive effect. The above authors have also reported that Chromolaena odorata as an excellent inhibitor for aluminium corrosion [80]. The environmentally friendly inhibitor could find possible applications in metal surface anodizing and surface coating in industries.

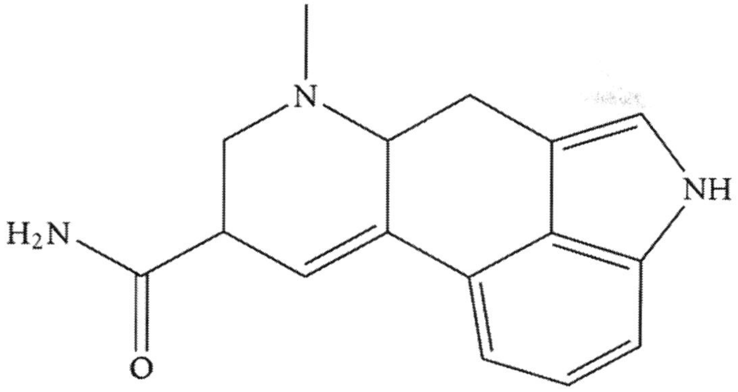

7-Methyl-4, 6.6a, 7, 8, 9-hexahydro-
indolo[4, 3-fg]quinoline-9-carboxamide

Figure 2: Structure of lysergic acid.

Soya Bean

It is rich in proteins, which are often good inhibitors in acidic media.

Most natural extracts constitute of oxygen- and nitrogen-containing compounds. Most of the oxygen-containing constituents of the extracts is a hydroxy aromatic compound, for example, tannins, pectins, flavonoids, steroids, and glycosides. Tannins are believed to form a passivating layer of tannates on the metallic surface. Similarly, it is postulated that a number of OH groups around the molecule lure them to form strong links with hydrogen and form complexes with metals. The complexes thus formed cause blockage of micro anodes and/or microanodes, which are generated on the metal surfaces when in contact with electrolytes, and, hence, retard subsequent dissolution of the metal.

Terminalia Catappa

The inhibitive and adsorption properties of ethanol extract of Terminalia catappa for the corrosion of mild steel in H_2SO_4 were investigated using weight loss, hydrogen evolution, and infrared methods of monitoring corrosion. The inhibition potential of ethanol extract of T.

catappa is attributed to the presence of saponin, tannin, phlobatin, anthraquinone, cardiac glycosides, flavanoid, terpene, and alkaloid in the extract. The adsorption of the inhibitor on mild steel surface is exothermic, spontaneous, and best described by Langmuir adsorption model [81] similar results were reported for Gnetum Africana [82].

Caffeic Acid

de Souza and Spinelli [83] studied the inhibitory action of caffeic acid as a green corrosion inhibitor for mild steel. The inhibitor effect of the naturally occurring biological molecule caffeic acid on the corrosion of mild steel in $0.1\,M$ H_2SO_4 was investigated by weight loss, potentiodynamic polarization, electrochemical impedance, and Raman spectroscopy. The different techniques confirmed the adsorption of caffeic acid onto the mild steel surface and consequently the inhibition of the corrosion process. Caffeic acid acts by decreasing the available cathodic reaction area and modifying the activation energy of the anodic reaction.

Gossypium Hirsutum

The corrosion inhibition properties of Gossypium hirsutum L leave extracts (GLE) and seed extracts (GSE) in $2\,M$ sodium hydroxide (NaOH) solutions were studied using chemical technique. Gossypium extracts inhibited the corrosion of aluminium in NaOH solution. The inhibition efficiency increased with increasing concentration of the extracts. The leave extract (GLE) was found to be more effective than the seed extract (GSE). The GLE gave 97% inhibition efficiency while the GSE gave 94% at the highest concentration [83].

It is found that ethanol extract of M. sapientum peels (banana) can be used as an inhibitor for mild steel corrosion. The inhibitor acts by being adsorbed on mild steel surface according to classical adsorption models of Langmuir and Frumkin adsorption isotherms. Adsorption characteristics of the inhibitor follow physical adsorption mechanism. It is found that temperature, pH, period of immersion, electrode potential, and concentration of the inhibitor basically control the inhibitive action of M. sapientum peels.

Carmine and Fast Green Dyes

The use of dyes such as azo compounds methyl yellow, methyl red, and methyl orange [84] as inhibitors for mild steel has been reported [85–87]. The inhibition action of carmine and fast green dyes on corrosion of mild steel in 0.5 M HCl was investigated using mass loss, polarization, and electrochemical impedance (EIS) methods. Fast green showed inhibition efficiency of 98% and carmine 92%. The inhibitors act as mixed type with predominant cathodic effect.

Corrosion inhibition of mild steel in acidic solution by the dye molecules can be explained on the basis of adsorption on the metal surface, due to the donor-acceptor interaction between π electrons of donor atoms N, O and aromatic rings of inhibitors, and the vacant d-orbitals of iron surface atoms [88, 89]. The fast green molecules possess electroactive nitrogen, oxygen atoms, and aromatic rings, favouring the adsorption while the carmine molecules possess electroactive oxygen atoms and electron rich paraquinanoid aromatic rings. In addition, the large and flat structure of the molecules occupies a large area of the substrate and thereby forming a protective coating. The inhibitors were adsorbed on the mild steel surface according to the Temkin adsorption isotherm (Figure 3).

(a)

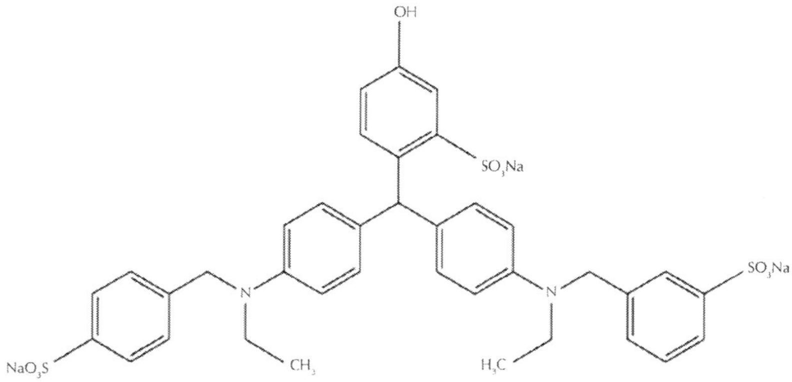

(b)

Figure 3: Structure of (a) carmine and (b) fast green.

Torres et al. [90] studied the effects of aqueous extracts of spent coffee grounds on the corrosion of carbon steel in a 1 mol L^{-1} HCl. Two methods of extraction were studied: decoction and infusion. The inhibition efficiency of C-steel in 1 mol L^{-1} HCl increased as the extract concentration and temperature increased. The coffee extracts acted as a mixed-type inhibitor with predominant cathodic effectiveness. In this study, the adsorption process of components of spent coffee grounds extracts obeyed the Langmuir adsorption isotherm. The chlorogenic acids isolated do not appear to be the active ingredient.

Biocorrosion and Prevention by Green Inhibitors

Biocorrosion relates to the presence of micro organisms that adhere to different industrial surfaces and damage the metal. Bacterial cells encase themselves in a hydrated matrix of polysaccharides and protein and form a slimy layer known as biofilm. The biofilm is a gel consisting of approximately 95% water, microbial metabolic products like enzymes, extracellular polymeric substances, organic and inorganic acids, and also volatile compounds such as ammonia or hydrogen sulphide and inorganic detritus [90–92]. Extracellular polymeric substances play a crucial role in biofilm development. Inhibition of biofilm formation is the simplest way of biocorrosion prevention. Use of naturally produced compounds such as plant extracts could be used as effective biocides [34].

SOL-GEL COATINGS

In recent years, the sol-gel coatings doped with inhibitors developed to replace chromate conversion coatings show real promise [93]. Results show that the corrosion resistance of the sol-gel coatings containing $CeCl_3$proves to be better than that of the pure and MBT-added sol-gel coatings by the electrochemical methods. However, unlike chromium, silane-based sol-gel coatings mainly act as physical barrier rather than form chemical bond with substrate. Inhibitors are necessary to release in the coating film to slow the corrosion process through self-healing effect [57, 89, 94–96]. Among the inhibitors, rare-earth elements are generally considered to be effective and nontoxic in sol-gel coatings. Additionally, some organic inhibitors, especially heterocyclic compounds, are effective as slowly released inhibitors in sol-gel coating [97, 98]. Andreeva et al. suggested self-healing anticorrosion coatings based on pH [99, 100]. The approach to prevention of corrosion propagation on metal surfaces achieving the self-healing effect is based on suppression of accompanying physicochemical reactions. The corrosion processes are followed by changes of the pH value in the corrosive area and metal degradation. Self-healing or self-curing of the areas damaged by corrosion can be performed by three mechanisms: pH neutralization, passivation of the damaged metal surface by inhibitors entrapped between polyelectrolyte layers, and repair of the coating. The corrosion inhibitor incorporated as a component of the layer-by-layer film into the protective coating is responsible for the most effective mechanism of corrosion suppression. Quinolines are environmentally friendly corrosion inhibitors that are attracting more and more attention as alternatives to the harmful chromates.

Recent awareness of the corrosion inhibiting abilities of tannins, alkaloids, organic and amino acids, as well as organic dyes has resulted in sustained interest on the corrosion inhibiting properties of natural products of plant origin. Such investigation is of much importance because in addition to being environmentally friendly and ecologically acceptable, plant products are inexpensive, readily available, and renewable sources of materials. Although a number of insightful papers have been devoted to corrosion inhibition by plant extracts, reports on the detailed mechanisms of the adsorption process are still scarce. The drawback of most reports on plant extracts as corrosion inhibitors is that the active ingredient has not been identified.

In recent years, sol-gel coatings doped with green inhibitors show real promise for corrosion protection of a variety of metals and alloys.

COMPUTATIONAL MODELING FOR CORROSION

Simulation is a prognostic computational tool for complex scientific and engineering problems. The simplest simulation methods have been used for decades, but, with the increase in computational memory and speed simulation, have become the prevalent tool for analysis [101–103]. Simulation turns probability models into statistics problems where the results can be analyzed using standard statistical methods. The challenge of a simulation is to implement a procedure that efficiently captures the desired model characteristics. Often the goal of probability computations is the evaluation of high reliability. In fact, computation of high reliabilities itself is an ongoing research concern. Hence, there is no one way in which to do the computation. Monte Carlo simulation is the traditional and powerful method if computational complexity and time are not limiting. The Box-Muller method is also well known. A variety of techniques have been developed to reduce the number of simulations without compromising accuracy.

The study of corrosion involves the study of the chemical, physical, metallurgical, and mechanical properties of materials as it is a synergistic phenomenon in which the environment is as equally important as the materials involved. Computer modeling techniques can handle the study of complex systems such as corrosion and thus are appropriate and powerful tools to study the mechanism of action of corrosion and its inhibitors.

In the recent past, computer modeling techniques have been successfully applied to corrosion problems as summarized in review articles by Zamani et al. [104] and Munn [105]. The application of computer modeling techniques to corrosion ˌsystems requires an understanding of the physical phenomenon of corrosion and the mathematics which govern the corrosion process. In addition, knowledge of the numerical procedures which are the basis of computer modeling techniques is essential for accurate computational analyses. In addition, validation of the computer analysis results with experimental data is mandatory. Without a reasonably accurate description of the damage process at a scale that is pertinent to the desired application, probabilistic computations have minimal value for prognosis and life-cycle assessment.

For corrosion modeling, the materials characterization depends on the orientation of the material. Figure 4 is a composite of three optical micrographs of the perpendicular faces of a typical specimen of 7075-T6 aluminum alloy, where LT, LS, and TS are the rolling, long-transverse, and short-transverse planes, respectively. Visually, there is a difference in the three surfaces, and the variability in the location, size, and density of the particles is apparent. Thus, for eg when modeling for aircraft wings, the LS surface is the most significant surface to characterize because it is the surface in fastener holes subjected to high-stress loading.

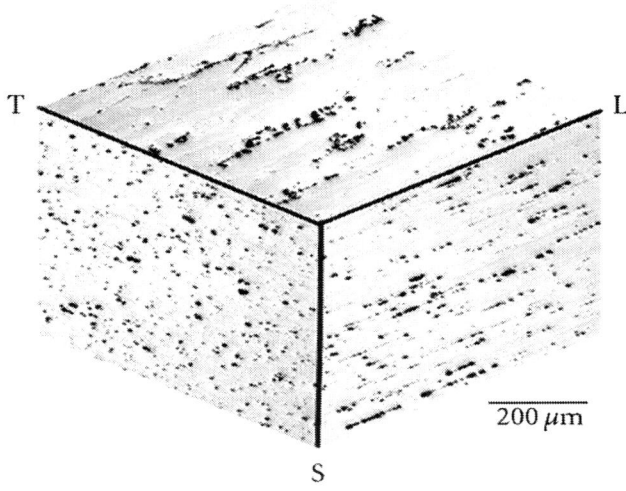

Figure 4: Three optical micrographs of the perpendicular faces of a typical specimen of 7075-T6 aluminum alloy.

Some Examples of Computational Modeling in Corrosion Inhibition

Tryptophan

According to the description of frontier orbital theory, HOMO is often associated with the electron donating ability of an inhibitor molecule. High EHOMO values indicate that the molecule has a tendency to donate electrons to the metal with unoccupied molecule orbitals. ELUMO indicates the ability of the molecules to accept electrons. The lower value of ELUMO is the easier acceptance of electrons from metal surface. The gap between the LUMO and HOMO energy levels of the inhibitor molecules is another important index, and the low absolute values of the energy band gap (DE = ELUMO − EHOMO) means good inhibition efficiency. Studies indicated that L-tryptophan has high value of EHOMO and low value of ELUMO with low-energy band gap. Adsorption energy calculated for the adsorption of L-tryptophan on Fe surface in the presence of water molecules equals −29.5 kJ mol^{-1}, which implies that the interaction between L-tryptophan molecule and Fe surface is strong [105, 106]. Molecule dynamics simulation results showed that L-tryptophan molecules assumed a nearly flat orientation with respect to the Fe (1 1 0) surface. The calculated adsorption energy between a L-tryptophan molecule and Fe surface is −29.5 kJmol^{-1} .

The optimized molecule structure, the highest occupied molecule orbitals, the lowest unoccupied molecule orbital, and the charge distribution of L-tryptophan molecule using DFT functional (B3LYP/6-311*G) are shown in Figure 5. The figure shows that in L-tryptophan molecule, C5, C12, C13, C14, C15, N7, N10, O2, and O4 carry more negative charges, while C8 and C6 carry more positive charges.

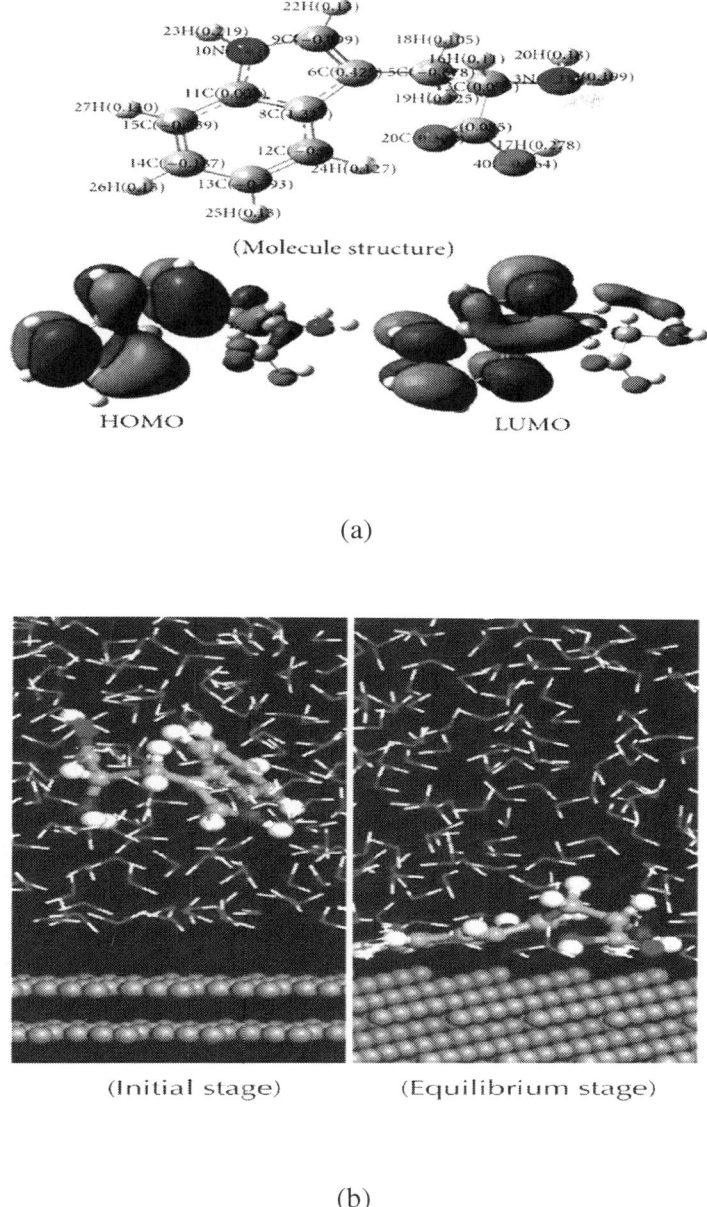

(Molecule structure)

HOMO   LUMO

(a)

(Initial stage)   (Equilibrium stage)

(b)

Figure 5: (a) Optimised molecule structure and charge density distribution of L-tryptophan. (b) L-tryptophan adsorbed on Fe surface in water solution.

This means that C5, C12, C13, C14, C15, N7, N10, O2, and O4 are the negative charge centers, which can offer electrons to the Fe atoms to form coordinate bond, and C8 and C6 are the positive charge centers, which can accept electrons from orbital of Fe atoms to form feedback bond. The optimized structure is in accordance with the fact that excellent corrosion inhibitors cannot only offer electrons to unoccupied orbital of the metal, but also accept free electrons from the metal. Therefore, it can be inferred that indole ring, nitrogen, and oxygen atoms are the possible active adsorption sites.

Presuel-Moreno et al. [107] modeled the chemical throwing power of an Al-Co-Ce metallic coating under thin electrolyte films representative of atmospheric conditions. An Al-Co-Ce alloy coating was developed for an AA2024-T3 substrate that can serve as barrier, sacrificial anode, and reservoir to supply soluble inhibitor ions to protect any defect sites or simulated scratches exposing the substrate. The model calculates the time necessary to accumulate Ce +3 and Co +2 inhibitors over the scratch when released from the Al-Co-Ce coating under different conditions such as the pH-dependent passive dissolution rate of an Al-Co-Ce alloy to define the inhibitor release flux. Transport by both electromigration and diffusion was considered. The effects of scratch size, initial pH, chloride concentration, and electrochemical kinetics of the material involved were studied. Studies indicated that sufficient accumulation of the released inhibitor (i.e., the Ce +3 concentration surpassed the critical inhibitor concentration over AA2024-T3 scratches) was achieved within a few hours (e.g., ~4 h for scratches of S = 1500 μm) when the initial solution pH was 6 and the coating was adjacent to the AA2024-T3.

Pradip and Rai [108] modeled design of phosphonic-acid-based corrosion inhibitors using a force field approach.

Piperidine and Derivatives

Khaled and Amin [109] studied the adsorption and corrosion inhibition behaviour of four selected piperidine derivatives, namely, piperidine (pip), 2-methylpiperidine (2mp), 3-methylpiperidine (3mp), and 4-methylpiperidine (4mp) at nickel in 1.0 M HNO_3 solution computationally by the molecular dynamics simulation and quantum chemical calculations and electrochemically by Tafel and impedance methods. The molecular dynamics (MD) simulations were performed using the commercial software MS Modeling from Accelrys using the amorphous cell module to create solvent piperidines cells on the nickel substrate. The behaviour of the inhibitors on the surface was studied using molecular dynamics simulations, and the condensed phase optimized molecular potentials for atomistic simulation studies (COMPASS) force field. COMPASS is an ab initio powerful force field which supports atomistic simulations of condensed phase materials [102]. Molecular simulation studies were applied to optimize the adsorption structures of piperidine derivatives. The nickel/inhibitor/solvent interfaces were simulated, and the charges on the inhibitor molecules as well as their structural parameters were calculated in the presence of solvent effects. Quantum chemical calculations based on the ab initio method were performed to determine the relationship between the molecular structure of piperidines and their inhibition efficiency. Results obtained from Tafel and impedance methods are in good agreement and confirm theoretical studies.

Khaled and Amin [110] also conducted studies on the molecular dynamics simulation on the corrosion inhibition of aluminum in molar hydrochloric acid using some imidazole derivatives. They also adapted Monte Carlo simulations technique incorporating molecular mechanics and dynamics to simulate the adsorption of methionine derivatives, namely, L-methionine, L-methionine sulphoxide, and L-methionine sulphone on iron (110) surface in 0.5 M sulphuric acid. Results show that methionine derivatives have a very good inhibitive effect for corrosion of mild steel in 0.5 M sulphuric acid solution.

Aniline and Its Derivatives

The inhibiting action of aniline and its derivatives on the corrosion of copper in hydrochloric acid has been investigated by Henriquez et al. [39], with emphasis on the role of substituents. With this purpose five different anilines were selected: aniline, p-chloroaniline, p-nitro aniline, p-methoxy, and p-methylaniline. A theoretical study using molecular mechanic and ab initio Hartree Fock methods, to model the adsorption of aniline on copper (100) showed results in good agreement with the experimental data. Aniline adsorbs parallel to the copper surface, showing no preference for a specific adsorption site. On the other hand, from ab initio Hartree Fock calculations, adsorption energy between 2 kcal/mol and 5 kcal/mol is obtained, which is close to the experimental value, confirming that the adsorption of aniline on the metal substrate is rather weak. In view of these results, the orientation of the aniline molecule with respect to the copper surface is considered to be the dominant effect. Mechanic molecular calculations were carried out using the Insight II, a comprehensive graphic molecular modeling program, to obtain configurations of minimum energy.

ACKNOWLEDGMENTS

The encouragement and cooperation received from Dr. Upadhya, Director, NAL, Bangalore, Dr. Ranjan Moodithaya, Head, KTMD, and Dr. K.S. Rajam, Head, SED are gratefully acknowledged. Patents used in the paper are: (1) US Patent 5951747—Non-chromate corrosion inhibitors for aluminum alloys; (2) United States Patent 5286357—Corrosion sensors; (3) WO/2002/008345—CORROSION INHIBITORS; and (4) British patent, 2327,1895.

REFERENCES

1. M. Bouklah, B. Hammouti, T. Benhadda, and M. Benkadour, "Thiophene derivatives as effective inhibitors for the corrosion of steel in 0.5 M H_2SO_4," Journal of Applied Electrochemistry, vol. 35, no. 11, pp. 1095–1101, 2005.

2. A. S. Fouda, A. A. Al-Sarawy, and E. E. El-Katori, "Pyrazolone derivatives as corrosion inhibitors for C-steel HCl solution," Desalination, vol. 201, pp. 1–13, 2006.

3. A. Fiala, A. Chibani, A. Darchen, A. Boulkamh, and K. Djebbar, "Investigations of the inhibition of copper corrosion in nitric acid solutions by ketene dithioacetal derivatives," Applied Surface Science, vol. 253, no. 24, pp. 9347–9356, 2007.

4. U. R. Evans, The Corrosion and Oxidation of Metals, Hodder Arnold, 1976.

5. O. K. Abiola, N. C. Oforka, E. E. Ebenso, and N. M. Nwinuka, "Eco-friendly corrosion inhibitors: The inhibitive action of Delonix Regia extract for the corrosion of aluminium in acidic

media," Anti-Corrosion Methods and Materials, vol. 54, no. 4, pp. 219–224, 2007.

6. M. Kliskic, J. Radoservic, S. Gudic, and V. Katalinic, "Aqueous extract of Rosmarinus officinalis L. as inhibitor of Al-Mg alloy corrosion in chloride solution," Journal of Applied Electrochemistry, vol. 30, no. 7, pp. 823–830, 2000.

7. A. Y. El-Etre, "Natural honey as corrosion inhibitor for metals and alloys. I. Copper in neutral aqueous solution," Corrosion Science, vol. 40, no. 11, pp. 1845–1850, 1998.

8. A. Y. El-Etre, "Inhibition of aluminum corrosion using Opuntia extract," Corrosion Science, vol. 45, no. 11, pp. 2485–2495, 2003.

9. A. Y. El-Etre, "Khillah extract as inhibitor for acid corrosion of SX 316 steel," Applied Surface Science, vol. 252, no. 24, pp. 8521–8525, 2006.

10. E. E. Ebenso, U. J. Ibok, U. J. Ekpe et al., "Corrosion inhibition studies of some plant extracts on aluminium in acidic medium," Transactions of the SAEST, vol. 39, no. 4, pp. 117–123, 2004.

11. E. E. Ebenso and U. J. Ekpe, "Kinetic study of corrosion and corrosion inhibition of mild steel in H_2SO_4 using Parica papaya leaves extract," West African Journal of Biological and Applied Chemistry, vol. 41, pp. 21–27, 1996.

12. U. J. Ekpe, E. E. Ebenso, and U. J. Ibok, "Inhibitory action of Azadirachta indica leaves extract on the corrosion of mild steel in H_2SO_4," West African Journal of Biological and Applied Chemistry, vol. 37, pp. 13–30, 1994.

13. F. Zucchi and I. H. Omar, "Plant extracts as corrosion inhibitors of mild steel in HCl solutions," Surface Technology, vol. 24, no. 4, pp. 391–399, 1985.

14. S. A. Umoren, O. Ogbobe, I. O. Igwe, and E. E. Ebenso, "Inhibition of mild steel corrosion in acidic medium using synthetic and naturally occurring polymers and synergistic halide additives," Corrosion Science, vol. 50, no. 7, pp. 1998–2006, 2008.

15. Y. J. Yee, Green inhibitors for corrosion control: a Study on the inhibitive effects of extracts of honey and rosmarinus officinalis L. (Rosemary), M.S. thesis, University of Manchester, Institute of Science and Technology, 2004.

16. J. C. Chalchat, R. P. Garry, A. Michet, B. Benjilali, and J. L. Chabart, "Essential oils of Rosemary (Rosmarinus officinalis L.). The chemical composition of oils of various origins (Morocco, Spain, France)," Journal of Essential Oil Research, vol. 5, no. 6, pp. 613–618, 1993.

17. E. El Ouariachi, J. Paolini, M. Bouklah et al., "Adsorption properties of Rosmarinus of ficinalis oil as green corrosion inhibitors on C38 steel in 0.5 M H_2SO_4," Acta Metallurgica Sinica, vol. 23, no. 1, pp. 13–20, 2010.

18. A. O. Odiongenyi, S. A. Odoemelam, and N. O. Eddy, "Corrosion inhibition and adsorption properties of ethanol extract of Vernonia Amygdalina for the corrosion of mild steel in H_2SO_4," Portugaliae Electrochimica Acta, vol. 27, no. 1, pp. 33–45, 2009.

19. S. A. Umoren and E. E. Ebenso, "Studies of the anti-corrosive effect of Raphia hookeri exudate gum-halide mixtures for aluminium corrosion in acidic medium," Pigment and Resin Technology, vol. 37, no. 3, pp. 173–182, 2008.

20. M. Abdallah, "Guar gum as corrosion inhibitor for carbon steel in sulphuric acid solutions," Portugaliae Electrochimica Acta, vol. 22, pp. 161–175, 2004.

21. P. C. Okafor, U. J. Ekpe, E. E. Ebenso, E. M. Umoren, and K. E. Leizou, "Inhibition of mild steel corrosion in acidic medium by Allium sativum extracts," Bulletin of Electrochemistry, vol. 21, no. 8, pp. 347–352, 2005.

22. P. C. Okafor and E. E. Ebenso, "Inhibitive action of Carica papaya extracts on the corrosion of mild steel in acidic media and their adsorption characteristics," Pigment and Resin Technology, vol. 36, no. 3, pp. 134–140, 2007.

23. P. C. Okafor, V. I. Osabor, and E. E. Ebenso, "Eco-friendly corrosion inhibitors: Inhibitive action of ethanol extracts of Garcinia kola for the corrosion of mild steel in H_2SO_4 solutions," Pigment and Resin Technology, vol. 36, no. 5, pp. 299–305, 2007.

24. P. C. Okafor, M. E. Ikpi, I. E. Uwah, E. E. Ebenso, U. J. Ekpe, and S. A. Umoren, "Inhibitory action of Phyllanthus amarus extracts on the corrosion of mild steel in acidic media," Corrosion Science, vol. 50, no. 8, pp. 2310–2317, 2008.

25. A. Y. El-Etre and M. Abdallah, "Natural honey as corrosion inhibitor for metals and alloys. II. C-steel in high saline water," Corrosion Science, vol. 42, no. 4, pp. 731–738, 2000.

26. A. Chetouani, B. Hammouti, and M. Benkaddour, "Corrosion inhibition of iron in hydrochloric acid solution by jojoba oil," Pigment and Resin Technology, vol. 33, no. 1, pp. 26–31, 2004.

27. A. Bouyanzer and B. Hammouti, "A study of anti-corrosive effects of Artemisia oil on steel," Pigment and Resin Technology, vol. 33, no. 5, pp. 287–292, 2004.

28. E. E. Oguzie, "Inhibition of acid corrosion of mild steel by Telfaria occidentalis extract," Pigment and Resin Technology, vol. 34, no. 6, pp. 321–326, 2005.

29. E. E. Oguzie, "Studies on the inhibitive effect of Occimum viridis extract on the acid corrosion of mild steel," Materials Chemistry and Physics, vol. 99, pp. 441–446, 2006.

30. E. E. Oguzie, "Corrosion inhibition of aluminium in acidic and alkaline media by Sansevieria trifasciataextract," Corrosion Science, vol. 49, no. 3, pp. 1527–1539, 2007.

31. M. A. Bendahou, M. B. E. Benadellah, and B. B. Hammouti, "A study of rosemary oil as a green corrosion inhibitor for steel in 2 M H_3PO_4," Pigment and Resin Technology, vol. 35, no. 2, pp. 95–100, 2006.

32. M. G. Sethuraman and P. B. Raja, "Corrosion inhibition of mild steel by Datura metel in acidic medium," Pigment and Resin Technology, vol. 34, no. 6, pp. 327–331, 2005.

33. N. O. Eddy, S. A. Odoemelam, and A. O. Odiongenyi, "Ethanol extract of musa species peels as a green corrosion inhibitor for mild steel: Kinetics, adsorption and thermodynamic considerations," Electronic Journal of Environmental, Agricultural and Food Chemistry, vol. 8, no. 4, pp. 243–255, 2009.

34. P. Deepa Rani and S. Selvaraj, "Inhibitive and adsorption properties of punica granatum extract on brass in acid media," Journal of Phytology, vol. 2, no. 11, pp. 58–64, 2010.

35. S. Rajendran, V. Ganga Sri, J. Arockiaselvi, and A. J. Amalraj, "Corrosion inhibition by plant extracts—an overview," Bulletin of Electrochemistry, vol. 21, no. 8, pp. 367–377, 2005.

36. K. Srivastava and P. Srivastava, "Studies on plant materials as corrosion inhibitors," British Corrosion Journal, vol. 16, no. 4, pp. 221–223, 1981.

37. R. M. Saleh, A. A. Ismail, and A. A. El Hosary, "corrosion inhibition by naturally occurring substances. vii. the effect of aqueous extracts of some leaves and fruit peels on the corrosion of steel, Al, Zn and Cu in acids," British Corrosion Journal, vol. 17, no. 3, pp. 131–135, 1982.

38. P. B. Raja and M. G. Sethuraman, "Natural products as corrosion inhibitor for metals in corrosive media—a review," Materials Letters, vol. 62, no. 1, pp. 113–116, 2008.

39. J. H. Henriquez-Román, M. Sancy, M. A. Páez et al., "The influence of aniline and its derivatives on the corrosion behaviour of copper in acid solution," Journal of Solid State Electrochemistry, vol. 9, no. 7, pp. 504–511, 2005.

40. A. A. El Hosary, R. M. Saleh, and A. M. Shams El Din, "Corrosion inhibition by naturally occurringsubstances-I. The effect of Hibiscus subdariffa (karkade) extract on the dissolution of Al and Zn," Corrosion Science, vol. 12, no. 12, pp. 897–904, 1972.

41. R. M. Saleh and A. M. Shams El Din, "Efficiency of organic acids and their anions in retarding the dissolution of aluminium," Corrosion Science, vol. 12, no. 9, pp. 689–697, 1972.

42. M. J. Sanghvi, S. K. Shuklan, A. N. Misra, M. R. Padh, and G. N. Mehta, "Inhibition of hydrochloric acid corrosion of mild steel by aid extracts of Embilica officianalis, Terminalia bellirica and Terminalia chebula," Bulletin of Electrochemistry, vol. 13, no. 8-9, pp. 358–361, 1997.

43. M. J. Shangvi, S. K. Shukla, A. N. Mishra, M. R. padh, and G. N. Mehta, "Corrosion inhibition of mild steel in hydrochloric acid by acid extracts of Sapindus Trifolianus, Acacia Concian and Trifla,"Transactions of the Metal Finishers Association of India, vol. 5, no. 3, pp. 143–147, 1996.

44. A. Chetouani and B. Hammouti, "Corrosion inhibition of iron in hydrochloric acid solutions by naturally henna," Bulletin of Electrochemistry, vol. 19, no. 1, pp. 23–25, 2003.

45. B. Muller, W. Klager, and G. Kubitzki, "Metal chelates of citric acid as corrosion inhibitors for zinc pigment," Corrosion Science, vol. 39, no. 8, pp. 1481–1485, 1997.

46. A. Bouyanzer, L. Majidi, and B. Hammouti, "Effect of eucalyptus oil on the corrosion of steel in 1M HCl," Bulletin of Electrochemistry, vol. 22, no. 7, pp. 321–324, 2006.

47. A. Y. El-Etre, "Natural onion juice as inhibitor for zinc corrosion," Bulletin of Electrochemistry, vol. 22, no. 2, pp. 75–80, 2006.

48. I. Radojcic, K. Berković, S. Kovač, and J. Vorkapić-Furač, "Natural honey and black radish juice as tin corrosion inhibitors," Corrosion Science, vol. 50, no. 5, pp. 1498–1504, 2008.

49. A. Y. El-Etre, M. Abdallah, and Z. E. El-Tantawy, "Corrosion inhibition of some metals using lawsonia extract," Corrosion Science, vol. 47, no. 2, pp. 385–395, 2005.

50. S. A. Umoren, I. B. Obot, and E. E. Ebenso, "Corrosion inhibition of aluminium using exudate gum from Pachylobus edulis in the presence of halide ions in HCl," E-Journal of Chemistry, vol. 5, no. 2, pp. 355–364, 2008.

51. N. O. Eddy and E. E. Ebenso, "Adsorption and inhibitive properties of ethanol extracts of Musa sapientum peels as a green corrosion inhibitor for mild steel in H_2SO_4," African Journal of Pure and Applied Chemistry, vol. 2, no. 6, pp. 046–054, 2008.

52. S. Lyon, "A natural way to stop corrosion," Nature, vol. 427, no. 406, p. 407, 2004.

53. E. E. Oguzie, K. L. Iyeh, and A. I. Onuchukwu, "Inhibition of mild steel corrosion in acidic media by aqueous extracts from Garcinia kola seed," Bulletin of Electrochemistry, vol. 22, no. 2, pp. 63–68, 2006.

54. R. M. Saldo, A. A. Ismail, and A. A. El Hosary, "Corrosion Inhibition by naturally occurring substances," British Corrosion Journal, vol. 17, no. 3, pp. 131–135, 1990.

55. E. E. Oguzie, "Corrosion inhibitive effect and adsorption bBehaviour of Hibiscus Sabdariffa extract on mild steel in acidic media," Portugaliae Electrochimica Acta, vol. 26, pp. 303–314, 2008.

56. E. E. Oguzie, "Corrosion inhibitive effect and adsorption behaviour of Hibiscus sabdariffa extract on mild steel in acidic media," Portugaliae Electrochimica Acta, vol. 26, no. 3, pp. 303–314, 2008.

57. H. W. Shi, F. C. Liu, E. H. Han, and M. C. Sun, "Investigation on a sol-gel coating containing inhibitors on 2024-T3 aluminum alloy," Chinese Journal of Aeronautics, vol. 19, pp. S106–S112, 2006.

58. G. D. Davis, "The Use of Extracts of Tobacco Plants as Corrosion Inhibitors,"http://www.electrochem.Org/dl/ma/202/pdfs/0340. PDF.

59. A. Y. El-Etre, "Inhibition of acid corrosion of aluminum using vanillin," Corrosion Science, vol. 43, no. 6, pp. 1031–1039, 2001.

60. W. A. Badawy, F. M. Allhara, and A. S. ElAzab, "Electrochemical behaviour and corrosion inhibition of Al, Al-6061 and Al-Cu in neutral aqueous solutions," Corrosion Science, vol. 41, no. 4, pp. 709–727, 1999.

61. B. Müller, "Amino and polyamino acids as corrosion inhibitors for aluminium and zinc pigments,"Pigment and Resin Technology, vol. 31, no. 2, pp. 84–87, 2002.

62. M. Kliskic, J. Radosevi, and S. Gudic, "Pyridine and its derivatives as inhibitors of aluminium corrosion in chloride solution," Journal of Applied Electrochemistry, vol. 27, no. 6, pp. 947–952, 1997.

63. R. Solmaz, G. Kardaş, B. Yazici, and M. Erbil, "Citric acid as natural corrosion inhibitor for aluminium protection," Corrosion Engineering Science and Technology, vol. 43, no. 2, pp. 186–191, 2008.

64. Z. Sibel, "The effects of benzoic acid in chloride solutions on the corrosion of iron and aluminum,"Turkish Journal of Chemistry, vol. 26, no. 3, pp. 403–408, 2002.

65. K. Berkovic, S. Kovac, and J.Vorkapic-Furac, "Natural compounds as environmentally friendly corrosion inhibitors of aluminium," Acta Alimentaria, vol. 33, no. 3, pp. 237–247, 2004.

66. Y. Tao, X. Zhang, and Z. Gu, "The inhibition of corrosion of aluminum in acid environment byin situ electrocoagulation of polybutadienoic acid," Wuhan University Journal of Natural Sciences, vol. 3, no. 2, pp. 221–225, 1998.

67. B. Müller and M. Kurfeß, "Saccharide und deren Derivate als Korrosionsinhibitoren für Aluminiumpigmente in wäßrigen Medien," Materials and Corrosion, vol. 44, no. 9, pp. 373–378, 2004.

68. G. O. Avwiri and F. O. Igho, "Inhibitive action of Vernonia amygdalina on the corrosion of aluminium alloys in acidic media," Materials Letters, vol. 57, no. 22-23, pp. 3705–3711, 2003.

69. S. Manish Kumar, K. Sudesh, R. Ratnani, and S. P. Mathur, "Corrosion inhibition of Aluminium by extracts of Prosopis cineraria in acidic media," Bulletin of Electrochemistry, vol. 22, no. 2, pp. 69–74, 2006.

70. A. A. Hossary, M. M. Gauish, and R. M. Saleh, "Corrosion inhibitor formulations from coal-tar distillation products for acid cleaning of steel in HCl," in Proceedings of the 2nd International Symposium on Industrial and Oriented Basic Electrochemistry, pp. 6–18, SAEST, CECRI, Madras, India, 1980.

71. R. A. Tupikov, Y. G. Dragunov, I. L. Kharina, and D. S. Zmienko, "Protection of carbon steels against atmospheric corrosion in a wet tropical climate using gas-plasma metallization with aluminum,"Protection of Metals, vol. 44, no. 7, pp. 673–682, 2008.

72. P. Arora, T. Jain, and S. P. Mathur, Chemistry, vol. 1, p. 766, 2005.

73. M. A. Quraishi, D. K. Yadav, and I. Ahamad, "Green approach to corrosion inhibition by black pepper extract in hydrochloric acid solution," Open Corrosion Journal, vol. 2, pp. 56–60, 2009.

74. N. Lahhit, A. Bouyanzer, J.-M. Desjobert, et al., "Fennel (Foeniculum vulgare) essential oil as green corrosion Inhibitor of carbon steel in hydrochloric acid solution," Portugaliae Electrochimica Acta, vol. 29, no. 2, pp. 127–138, 2011.

75. E. E. Oguzie, "Corrosion inhibition of mild steel in hydrochloric acid solution by methylene blue dye,"Materials Letters, vol. 59, no. 8-9, pp. 1076–1079, 2005.

76. A. Zarrouk, I. Warad, B. Hammouti, A. Dafali, S. S. Al-Deyab, and N. Benchat, "The effect of temperature on the corrosion of Cu/HNO$_3$ in the Presence of organic inhibitor: Part-2," International Journal of Electrochemical Science, vol. 5, no. 10, pp. 1516–1526, 2010.

77. K. P. Vinod Kumar, M. S. Narayanan Pillai, and G. Rexin Thusnavis, "Pericarp of the fruit of garcinia mangostana as corrosion inhibitor for mild steel in hydrochloric acid medium,"

Portugaliae Electrochimica Acta, vol. 28, no. 6, pp. 373–383, 2010.

78. H. A. Jung, B. N. Su, W. J. Keller, R. G. Mehta, and A. D. Kinghorn, "Antioxidant xanthones from the pericarp of Garcinia mangostana (Mangosteen)," Journal of Agricultural and Food Chemistry, vol. 54, no. 6, pp. 2077–2082, 2006.

79. I. B. Obot, N. O. Obi-Egbedi, S. A. Umoren, and E. E. Ebenso, "Synergistic and antagonistic effects of anions and ipomoea invulcrata as green corrosion inhibitor for aluminium dissolution in acidic medium," International Journal of Electrochemical Science, vol. 5, no. 7, pp. 994–1007, 2010.

80. I. B. Obot and N. O. Obi-Egbedi, "Ipomoea involcrata as an ecofriendly inhibitor for aluminium in alkaline medium," Portugaliae Electrochimica Acta, vol. 27, no. 4, pp. 517–524, 2009.

81. I. B. Obot and N. O. Obi-Egbedi , "An interesting and efficient green corrosion inhibitor for aluminium from extracts of Chlomolaena odorata L. in acidic solution," Journal of Applied Electrochemistry, vol. 40, no. 11, pp. 1977–1983, 2010.

82. N. O. Eddy, P. A. Ekwumemgbo, and P. A. P. Mamza, "Ethanol extract of Terminalia catappa as a green inhibitor for the corrosion of mild steel in H_2SO_4," Green Chemistry Letters and Reviews, vol. 2, no. 4, pp. 223–231, 2009.

83. F. A. de Souza and A. Spinelli, "Caffeic acid as a green corrosion inhibitor for mild steel," Corrosion Science, vol. 51, no. 3, pp. 642–649, 2008.

84. O. K. Abiola, J. O. E. Otaigbe, and O. J. Kio, "Gossipium hirsutum L. extracts as green corrosion inhibitor for aluminum in NaOH solution," Corrosion Science, vol. 51, no. 8, pp. 1879–1881, 2009.

85. F. Tirbonod and C. Fiaud, "Inhibition of the corrosion of aluminium alloys by organic dyes," Corrosion Science, vol. 18, no. 2, pp. 139–149, 1978.

86. J. D. Talati and J. M. Daraji, "Inhibition of corrosion of B26S aluminium in phosphoric acid by some azo dyes," Journal of the Indian Chemical Society, vol. 68, no. 2, pp. 67–72, 1991.

87. P. Gupta, R. S. Chaudhary, T. K. G. Namboodhiri, B. Prakash, and B. B. Prasad, "Effect of mixed inhibitors on dezincification and corrosion of 63/37 brass in 1% sulfuric acid," Corrosion, vol. 40, no. 1, pp. 33–36, 1984.

88. P. B. Tandel and B. N. Oza, "Performance of some dyestuffs as inhibitors during corrosion of mild-steel in binary acid mixtures (HCl + HNO^3)," Journal of the Electrochemical Society of India, vol. 49, pp. 49–128, 2000.

89. M. L. Zheludkevich, R. Serra, M. F. Montemor, and M. G. S. Ferreira, "Oxide nanoparticle reservoirs for storage and prolonged release of the corrosion inhibitors," Electrochemistry Communications, vol. 7, no. 8, pp. 836–840, 2005.

90. V. V. Torres, R. S. Amado, C. Faia de Sá, et al., "Inhibitory action of aqueous coffee ground extracts on the corrosion of carbon steel in HCl solution," Corrosion Science, vol. 53, no. 7, pp. 2385–2392, 2011.

91. G. G. Geesey, "Microbial exopolymers: ecological and econimic considerations," American Society Microbiology News, vol. 48, pp. 9–14, 1982.

92. I. B. Beech and C. C. Gaylarde, "Recent advances in the study of biocorrosion—an overview," Revista de Microbiologia, vol. 30, no. 3, pp. 177–190, 1999.

93. P. S. Guiamet and S. G. Gomez De Saravia, "Laboratory studies of biocorrosion control using traditional and environmentally friendly biocides: an overview," Latin American Applied Research, vol. 35, no. 4, pp. 295–300, 2005.

94. M. F. Montemor, W. Trabelsi, M. Zheludevich, and M. G. S. Ferreira, "Modification of bis-silane solutions with rare-earth cations for improved corrosion protection of galvanized steel substrates,"Progress in Organic Coatings, vol. 57, no. 1, pp. 67–77, 2006.

95. A. Pepe, M. Aparicio, S. Ceré, and A. Durán, "Preparation and characterization of cerium doped silica sol-gel coatings on glass and aluminum substrates," Journal of Non-Crystalline Solids, vol. 348, pp. 162–171, 2004.

96. X. W. Yu, C. N. Cao, Z .M. Yao, Z. Derui, and Y. Zhongda, "Corrosion behavior of rare earth metal (REM) conversion coatings

on aluminum alloy LY12," Materials Science and Engineering A, vol. 284, no. 1-2, pp. 56–63, 2000.

97. A. N. Khramov, N. N. Voevodin, V. N. Balbyshev, and M. S. Donley, "Hybrid organo-ceramic corrosion protection coatings with encapsulated organic corrosion inhibitors," Thin Solid Films, vol. 447-448, pp. 549–557, 2004.

98. E. M. Sherif and S. M. Park, "Effects of 1,4-naphthoquinone on aluminum corrosion in 0.50 M sodium chloride solutions," Electrochimica Acta, vol. 51, no. 7, pp. 1313–1321, 2006.

99. M. S. Donley, R. A. Mantz, A. N. Khramov, V. N. Balbyshev, L. S. Kasten, and D. J. Gaspar, "The self-assembled nanophase particle (SNAP) process: a nanoscience approach to coatings," Progress in Organic Coatings, vol. 47, no. 3-4, pp. 401–415, 2003.

100. D. V. Andreeva, D. Fix, H. Möhwald, and D. G. Shchukin, "Self-healing anticorrosion coatings based on pH-sensitive polyelectrolyte/inhibitor sandwichlike nanostructures," Advanced Materials, vol. 20, no. 14, pp. 2789–2794, 2008.

101. M. L. Zheludkevich, D. G. Shchukin, K. A. Yasakau, H. Möhwald, and M. G. S. Ferreira, "Anticorrosion coatings with self-healing effect based on nanocontainers impregnated with corrosion inhibitor," Chemistry of Materials, vol. 19, no. 3, pp. 402–411, 2007.

102. V. DeGiorgi, "Corrosion basics and computer modeling," in Industrial Applications of the BEM, chapter 2, pp. 47–79, 1986.

103. H. Sun, P. Ren, and J. R. Fried, "The compass force field: parameterization and validation for polyphosphazenes," Computational and Theoretical Polymer, vol. 8, pp. 229–246, 1998.

104. N. G. Zamani, J. F. Porter, and A. A. Mufti, "Survey of computational efforts in the field of corrosion engineering," International Journal for Numerical Methods in Engineering, vol. 23, no. 7, pp. 1295–1311, 1986.

105. R. S. Munn, "A review of the development of Computational Corrosion analysis for special corrosion modelling through its maturity in the Mid 1980's-Computer modelling in Corrosion," ASTM STP 1154 American Study for Testing and Materials, Philadelphia, pp 215-228, 1991.

106. K. F. Khaled, "Molecular simulation, quantum chemical calculations and electrochemical studies for inhibition of mild steel by triazoles," Electrochimica Acta, vol. 53, no. 9, pp. 3484–3492, 2008.

107. F. J. Presuel-Moreno, H. Wang, M. A. Jakab, R. G. Kelly, and J. R. Scully, "Computational modeling of active corrosion inhibitor release from an Al-Co-Ce metallic coating," Journal of the Electrochemical Society, vol. 153, no. 11, Article ID 002611JES, pp. B486–B498, 2006.

108. Pradip and B. Rai, "Design of tailor-made surfactants for industrial applications using a molecular modelling approach," Colloids and Surfaces A, vol. 205, no. 1-2, pp. 139–148, 2002.

109. K. F. Khaled and M. A. Amin, "Computational and electrochemical investigation for corrosion inhibition of nickel in molar nitric acid by piperidines," Journal of Applied Electrochemistry, vol. 38, no. 11, pp. 1609–1621, 2008.

110. K. F. Khaled and M. A. Amin, "Electrochemical and molecular dynamics simulation studies on the corrosion inhibition of aluminum in molar hydrochloric acid using some imidazole derivatives," Journal of Applied Electrochemistry, vol. 39, no. 12, pp. 2553–2568, 2009.

Corrosion and Corrosion Inhibition of High Strength Low Alloy Steel in 2.0 M Sulfuric Acid Solutions by 3-Amino-1,2,3-Triazole as a Corrosion Inhibitor

El-Sayed M. Sherif[1, 2], Adel Taha Abbas[1], D. Gopi[3, 4], and A. M. El-Shamy[2]

[1]Mechanical Engineering Department, College of Engineering, King Saud University, P.O. Box 800, Riyadh 11421, Saudi Arabia

[2]Electrochemistry and Corrosion Laboratory, Department of Physical Chemistry, National Research Centre (NRC), Dokki, Cairo 12622, Egypt

[3]Department of Chemistry, Periyar University, Salem 636 011, Tamil Nadu, India

[4]Centre for Nanoscience and Nanotechnology, Periyar University, Salem 636 011, Tamil Nadu, India

ABSTRACT

The corrosion and corrosion inhibition of high strength low alloy (HSLA) steel after 10 min and 60 min immersion in 2.0 M H_2SO_4 solution by 3-amino-1,2,4-triazole (ATA) were reported. Several electrochemical techniques along with scanning electron microscopy (SEM) and energy dispersive X-ray (EDS) were employed. Electrochemical impedance spectroscopy indicated that the increase of immersion time from 10 min to 60 min significantly decreased both the solution and polarization resistance for the steel in the sulfuric acid solution. The increase of immersion time increased the anodic, cathodic, and corrosion currents, while it decreased the polarization resistance as indicated by the potentiodynamic polarization measurements. The addition of 1.0 mM ATA remarkably decreased the corrosion of the steel and this effect was found to increase with increasing its concentration to 5.0 mM. SEM and EDS investigations confirmed that the inhibition of the HSLA steel in the 2.0 M H_2SO_4 solutions is achieved via the adsorption of the ATA molecules onto the steel protecting its surface from being dissolved easily.

INTRODUCTION

High strength low alloy (HSLA) steels are designed to provide better mechanical properties and/or greater resistance to atmospheric corrosion than conventional carbon steels in the normal sense because they are designed to meet specific mechanical properties rather than a chemical composition [1]. HSLA steels have been widely used in many applications in industry; these include gun barrel, food sterilization, sintering of components from powders, hypersonic wind tunnels, power generation, and water jet cutting [1, 2]. There is a great economical incentive in developing methods and materials to alleviate corrosion, which comes only from a good understanding of the mechanisms and processes involved in this complex phenomenon [3, 4].

Acid solutions are widely used in many applications, such as acid pickling, industrial acid cleaning, acid descaling, and oil well acidizing [5]. Due to the corrosivity of acid solutions, corrosion inhibitors are commonly added to their solutions in order to reduce their aggressive attack on the structure to be protected [5–10]. Inhibitors are generally used in this process to control the metal dissolution as well as acid consumption [5]. It has been reported [11–16] that organic compounds containing polar groups including nitrogen, sulfur, and oxygen and heterocyclic compounds with polar functional groups and conjugated double bonds have been known to be good corrosion inhibitors. The inhibiting action of these compounds is usually attributed to their interactions with the metal via their adsorption onto the surface. However, the adsorption of an inhibitor on a metal surface depends on the nature and the surface charge of the metal, the adsorption mode, its chemical structure, and the type of the electrolyte solution [17].

In the present work, we reported the effect of immersion time, namely, 10 min and 60 min, on the electrochemical corrosion behavior of the HSLA steel in 2.0 M sulfuric acid solutions. The effect of adding different concentrations of 3-amino-1,2,4-triazole (ATA) on the inhibition of this steel after the different exposure periods in the acid solution was also reported. The experimental part of this study was carried out using open-circuit potential (OCP), electrochemical impedance spectroscopy (EIS), and potentiodynamic polarization measurements. Characterization of the surface of the steel after its immersion in the acid solution alone and in the acid solution containing ATA molecules was performed using scanning electron microscopy (SEM) and energy dispersive X-ray (EDS) analyses.

EXPERIMENTAL DETAILS

Chemicals and Electrochemical Cell

3-Amino-1,2,4-triazole (ATA, Sigma-Aldrich, 95%), sulfuric acid (H_2SO_4, Merck, 96%), absolute ethanol (C_2H_5OH, Merck, 99.9%), and acetone (C_3H_6O, Merck, 99.0%) were used as received. The HSLA steel electrode with a square shape and surface dimensions of 1×1 cm was employed for the electrochemical tests. The chemical composition of the employed ultrahigh strength steel (in wt.%) was as follows: C = 0.0309, Si =0.177, Mn = 0.201, P = 0.007, Cr = 1.553, Mo =0.617, Ni = 3.208, Nb = 0.002, Al = 0.006, Cu =0.098, Co = 0.011, B = 0.001, V = 0.223, Sn =0.002, and N = 0.017, and the balance (~93.566) was Fe. A conventional electrochemical cell accommodating only 200 mL with a three-electrode configuration was used. The three electrodes were the steel rod, platinum foil, and an Ag/AgCl electrode (in 3.0 M KCl) and were used as working, counter, and reference electrodes, respectively. The working electrode for electrochemical measurements was prepared by attaching an insulated copper wire to one face of the sample using an aluminum conducting tape and cold mounted in resin. The surface of the steel electrode to be exposed to the solution was first ground successively with metallographic emery paper of increasing fineness of up to 600 grits and further with 5, 1, 0.5, and 0.3 μm alumina slurries (Buehler). The electrode was then washed with doubly distilled water, degreased with acetone, washed using doubly distilled water again, and finally dried with tissue paper.

Electrochemical Measurements

An Autolab Potentiostat (PGSTAT20 computer controlled) operated by the General Purpose Electrochemical Software (GPES) version 4.9 was used to perform the electrochemical experiments. The open-circuit potential (OCP) curves obtained for the steel electrode in sulfuric acid in the absence and presence of ATA were done for 60 min. The electrochemical impedance spectroscopy (EIS) tests were performed at corrosion potentials over a frequency range of 100 kHz to 100 mHz, with an AC wave of ±5 mV peak-to-peak overlaid on a DC bias potential, and the impedance data were collected using Powersine software at a rate of 10 points per decade change in frequency. The potentiodynamic polarization curves were obtained by scanning the potential in the forward direction from −1.0 V to −0.2 V versus Ag/AgCl at a scan rate of 0.001 V/s. Each experiment was carried out using fresh steel surface and new portion of the sulfuric acid solution in the absence and the presence of the ATA molecules. All electrochemical experiments were carried out at room temperature.

SEM and EDS Investigations

The SEM images were obtained by using a JEOL model JSM-6610LV (Japanese made) scanning electron microscope with an energy dispersive X-ray analyzer attached for acquiring the EDS analysis.

RESULTS AND DISCUSSION

Open-Circuit Potential (OCP) Measurements

Figure 1 shows the OCP curves of the steel electrode in aerated stagnant 2.0 M H_2SO_4 solutions in the absence (1) and presence (2) of 1.0 mM and (3) 5.0 mM ATA, respectively. It is seen that the potential of the steel in the sulfuric acid solution without inhibitor (curve 1) increased towards the more negative values in the first 300 s of the steel immersion as a result of the dissolution of a preoxide film via the aggressiveness action of the acid solution. The potential then shifted again in the less negative direction through another 300 s before stabilizing a slight shift in the same direction with time till the end of the run.

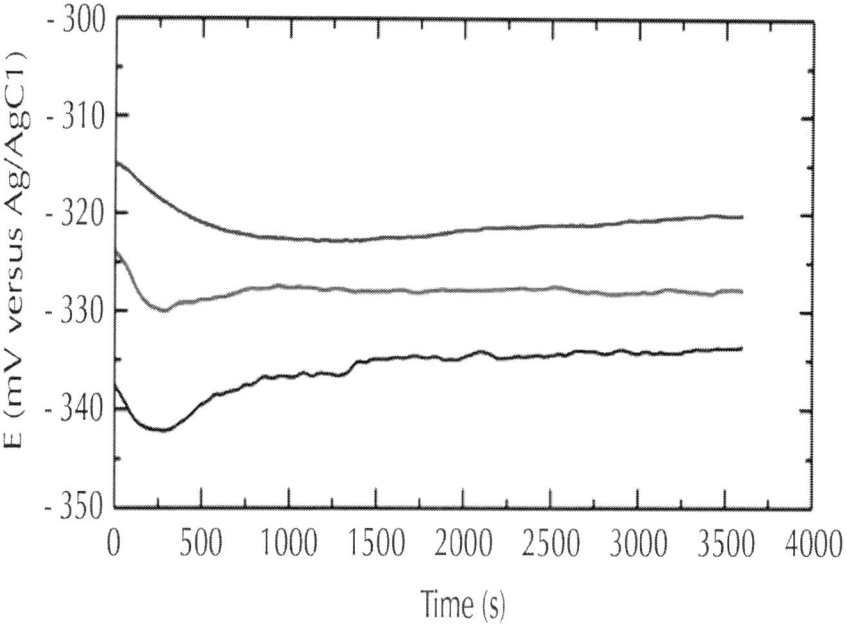

Figure 1: Change of the open-circuit potential versus time for the HSLA steel in 2.0 M H_2SO_4 solutions in the absence (1) and presence (2) of 1.0 mM ATA and (3) 5.0 mM ATA, respectively.

The addition of 1.0 mM ATA (curve 2) showed almost the same behavior but with a positive shift in the absolute potential, which indicates that ATA molecules at this concentration have an ability to decrease the severity of the acid solution. Increasing ATA concentration to 5.0 mM (curve 3) presented more positive shifts in the absolute potential of the steel for the whole time of the experiment. The change of OCP with time measurements thus indicated that the presence of ATA molecules decreases the aggressiveness action of H_2SO_4 by shifting its potential in the less negative direction and this effect was found to increase with increase of the concentration of ATA.

SEM And EDS Investigations

In order to investigate the effect of H_2SO_4 in the absence and presence of ATA molecules on the corrosion of the HSLA steel, SEM micrograph and EDS profile analysis were carried out. Figure 2 shows the SEM micrographs obtained for the HSLA steel after 24 h immersion in 2.0 M H_2SO_4 solutions, (a) without ATA and (b) with 5.0 mM ATA present, and (c) EDS profile analysis corresponding to the surface shown in Figure 2(b), respectively. The SEM micrograph, shown in Figure 2(a), shows a total deterioration for the surface of steel, which was due to the aggressiveness attack of the sulfuric acid solution. The EDS profile analysis taken for the steel at this condition [7] indicated that its surface has the main alloying elements, in addition to the presence of carbon, sulfur, and oxygen. The presence of C, O, and S was due to the effect of sulfuric acid solutions as well as the exposure of the steel surface to air after removing it from the acid solution.

(a)

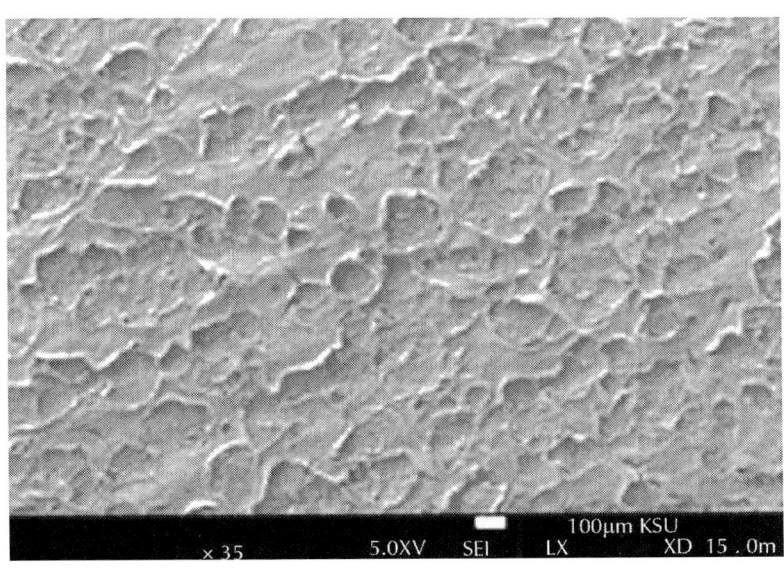

(b)

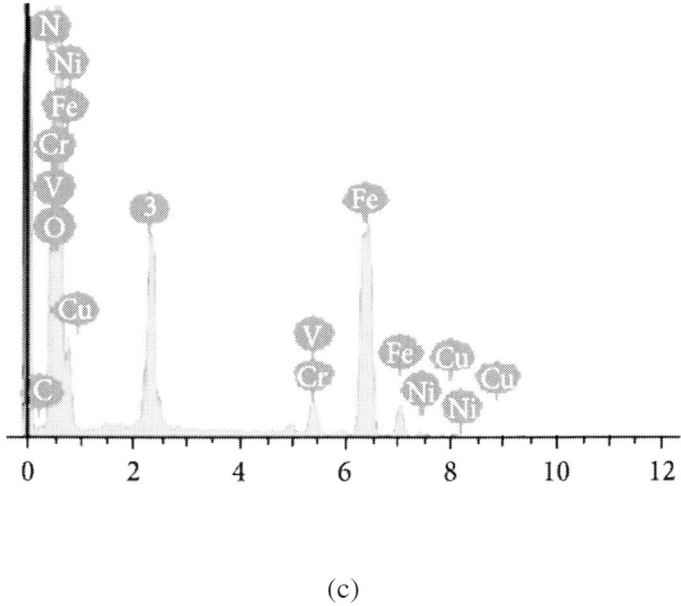

(c)

Figure 2: SEM micrographs obtained for the HSLA steel after 24 h immersion in 2.0 M H$_2$SO$_4$ solutions, (a) without ATA and (b) with 5.0 mM ATA present, and (c) EDS profile analysis corresponding to the surface shown in Figure 2(b), respectively.

On the other hand, the SEM micrograph shown in Figure 2(b) proved that the surface of the steel is covered with a homogeneous layer of the adsorbed ATA molecules. This was confirmed by the EDS profile analysis shown in Figure 2(c), where the atomic percentages of the elements found on the steel surface were 5.44% C, 0.93% N, 53.65% O, 7.34% S, 0.69% V, 2.49% Cr, 0.77% Ni, 0.12% Cu, and 28.58% Fe. The presence of nitrogen in the analysis confirms that the ATA molecules are included in the layer present on the surface. The presence of very high amount of Cr compared to that originally present in the steel, in addition to the very high percent of oxygen, indicates that the surface is also passivated through the formation of chromium oxide layer along with the adsorbed layer of the ATA molecules. Moreover, the presence of very low amounts of Fe and Ni reveals that the formed ATA layer is thick and is homogenously distributed on the surface. Another proof for the ability of the ATA molecules to inhibit the HSLA steel corrosion in the sulfuric acid test solution was the black color of the solution which did not have any ATA compound and was due to the severe dissolution of the steel, while the color of the acid solution containing 5.0 mM ATA was clear and unchanged even after 24 h of the steel immersion. It has been reported [9, 18–20] that the inhibition of metal corrosion by using similar compounds to ATA is achieved by the adsorption of their molecules onto the metal surface preventing it from being attacked by corrosive media.

Electrochemical Impedance Spectroscopy (EIS) Measurements

It has been reported [10, 20–22] that EIS is a powerful method in understanding the corrosion and corrosion inhibition for different metals and alloys in aggressive environments. In this work, we employed the EIS experiments to obtain the kinetic parameters for the steel/sulfuric acid solution interface after different exposure periods. In order to report the effect of immersion time on the corrosion of the HSLA steel in 2.0 M H_2SO_4 solution, the EIS measurements were conducted after (1) 10 min and (2) 60 min immersion and the Nyquist spectra were plotted, respectively, as shown in Figure 3.

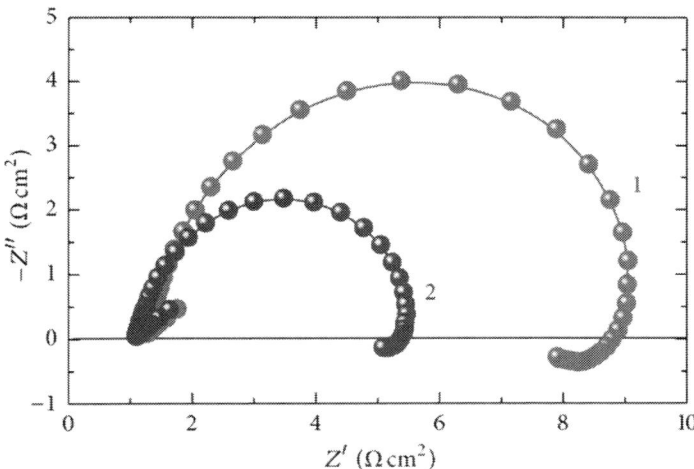

Figure 3: Nyquist plots for the HSLA steel after its immersion for (1) 10 min and (2) 60 min in the 2.0 M H_2SO_4 solutions, respectively.

It is clearly seen that the steel in the acid solution shows only one distorted semicircle whose diameter got smaller with the increase of immersion time from 10 min to 60 min. This indicates that the increase of the time of immersion increases the dissolution of steel in sulfuric acid solution through decreasing its corrosion resistance. This was confirmed by fitting the EIS data to the best equivalent circuit model, which is shown in Figure 4. This equivalent circuit model was also used to fit the EIS data obtained from studying the corrosion and corrosion inhibition of maraging steel in different sulfuric acid solutions [7, 9]. The parameters of the used circuit can be defined according to usual convention as follows: R_s represents the solution resistance, Q is the constant phase elements (CPEs), R_{p1} is the polarization resistance for the solution/steel interface and can be defined as the charge transfer resistance [23], R_{p2} is another polarization resistance for the corrosion product/steel interface, and L is the inductance. The values of these parameters are listed in Table 1. It is also seen from Table 1 that the values of R_s, R_{p1}, R_{p2}, and decrease, while the value of Q (CPEs) increases by the increase of immersion time from 10 min to 60 min for the HSLA steel in 2.0 M H_2SO_4 solutions. This is due to the corrosiveness action of the sulfuric acid solution that continuously attacks the surface of the steel and lowers its probability to develop oxide layers or corrosion products and that effect increases with increasing time of contact between the acid and the steel.

Table 1: Parameters obtained by fitting the EIS data with the equivalent circuit shown in Figure 4 for the HSLA steel in 2.0 M H_2SO_4 solutions

Solution	Parameter						
	$Rs/\Omega cm2$	Q		$Rp1/\Omega cm2$	$Rp2/\Omega cm2$	L/H	IE %
		$YQ/F cm-2$	n				
2.0 M H2SO4(10 min)	1.357	0.00145	0.80	6.766	3.204	15.347	—
+1.0 mM ATA (10 min)	1.491	0.00122	0.88	10.747	3.301	15.611	37.1
+5.0 mM ATA (10 min)	1.860	0.00103	0.92	14.000	3.977	16.92	51.7
2.0 M H2SO4(60 min)	1.280	0.00257	0.84	5.354	2.279	13.73	—
+1.0 mM ATA (60 min)	1.434	0.00187	0.87	8.993	3.267	15.716	40.5
+5.0 mM ATA (60 min)	1.657	0.00179	0.92	12.07	3.355	16.602	55.64

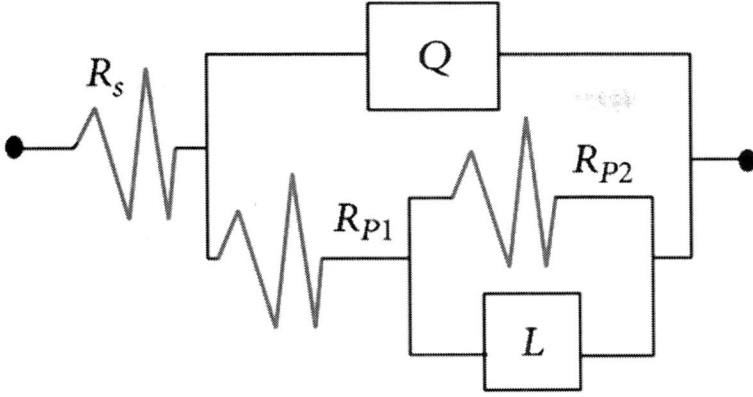

Figure 4: The equivalent circuit used to fit the experimental EIS data obtained for the HSLA steel after its immersion for different periods of time in 2.0 M sulfuric acid solution.

The EIS measurements were also employed to report the effect of ATA molecules on the inhibition of the HSLA steel in 2.0 M H_2SO_4 solutions. Figure 5 shows the Nyquist plots obtained for the steel after 10 min immersion in the acid solution that contains (1) 0.0, (2) 1.0, and (3) 5.0 mM ATA, respectively. Similar plots were also obtained for the HSLA steel after its immersion in the acid solution in the absence and presence of ATA for 60 min and the curves are shown in Figure 6. The equivalent circuit shown in Figure 4 was also used to represent the best fitting for the data presented in Figures 5 and 6. The EIS parameters obtained out of the fitted data as well as the values of the percentage of the inhibition efficiency (IE %) are listed in Table 1. The values of IE% were calculated according to the following equation [18]:

$$IE\% = \frac{R_P^{in} - R_P^{O}}{R_P^{in}},$$

(1)

where R_P^{in} and R_P^{O} are polarization resistance of the HSLA steel in the sulfuric acid solution in the presence and absence of ATA molecules, respectively.

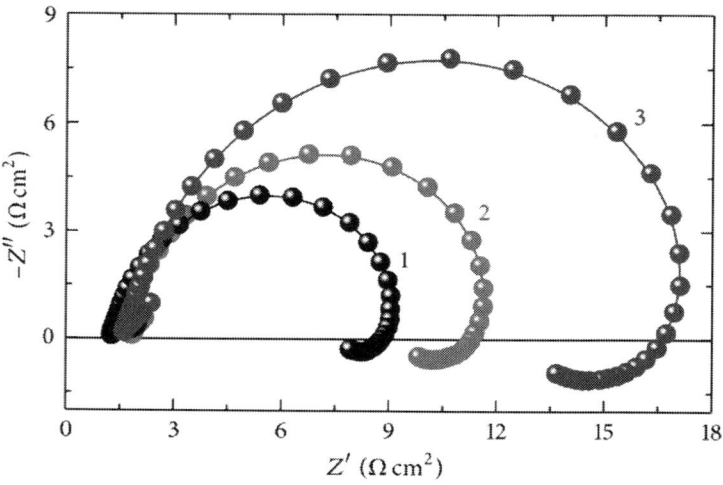

Figure 5: Nyquist plots for the HSLA steel after its immersion for 10 min in 2.0 M H$_2$SO$_4$solutions in (1) the absence and the presence of (2) 1 mM ATA and (3) 5 mM ATA, respectively.

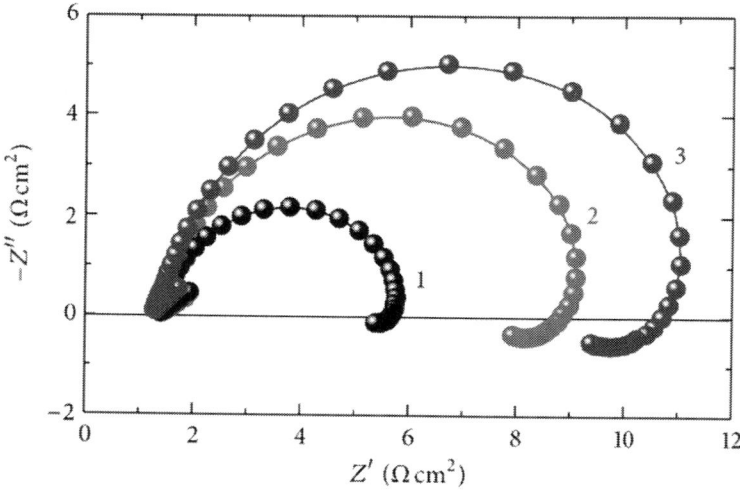

Figure 6: Nyquist plots for the HSLA steel after its immersion for 60 min in 2.0 M H$_2$SO$_4$solutions in (1) the absence and the presence of (2) 1 mM ATA and (3) 5 mM ATA, respectively.

It is clearly seen from Figure 5 that the addition of 1.0 mM ATA within the sulfuric acid solution increased the diameter of the semicircle. Increasing the concentration of ATA to 5.0 mM further increased the diameter of the semicircle obtained for the steel. This indicates that the presence of ATA and the increase of its concentration decrease the corrosion of the HSLA steel after 10 min immersion in 2.0 M H_2SO_4 solutions. It has been reported [18, 20, 24] that ATA molecules inhibit the corrosion via their adsorption onto the surface of metals (such as iron [20] and copper [18, 24]) and the ability of ATA as a corrosion inhibitor increases with the increase of its concentration. This agrees with the current results as indicated by the parameters recorded in Table 1, where the values of R_s, R_{p1}, R_{p2}, and L as well as IE% increased with the increase of ATA concentration.

It is obvious from Figure 6 for the steel after 60 min in the acid solution without and with ATA present that the presence of 1.0 mM ATA decreased the aggressiveness action of the sulfuric acid solution by increasing its corrosion resistance. This was revealed by increasing the diameter of the semicircle obtained for the steel in the presence of 1.0 mM ATA compared to its diameter in the blank solution. Further increasing the ATA concentration to 5.0 mM produced further increase in the diameter of the semicircle after 60 min. Table 1 also confirmed that the presence of ATA and the increase of its concentration increased the values of R_s, R_{p1}, R_{p2}, and L The increase of R_s, R_{p1}, and R_{p2}, and L in the presence of ATA and with the increase of its content indicates that ATA molecules have the ability to increase the solution and corrosion resistance of the HSLA steel surface and that effect increases with increasing ATA concentration in the acid solution. The values of CPEs with their n values close to unity represent double layer capacitors decreased in the presence of ATA and with the increase of its content, which was expected to cover up the charged surfaces [23]. Moreover, the values of IE% were found to increase not only with increasing concentration of ATA from 1.0 mM to 5.0 mM but also with increasing the immersion time as can be seen from Table 1. The outcome of the EIS experiments proves that ATA can be employed to mitigate the corrosion of the HSLA steel in 2.0 M H_2SO_4 solutions and its ability as a good corrosion inhibitor increases with the increase of its concentration as well as the increase of the time of immersion before measurement.

Potentiodynamic Polarization Measurements

The effect of increasing the immersion time on the dissolution of the HSLA steel in 2.0 M H_2SO_4 solutions was investigated using potentiodynamic polarization measurements. The potentiodynamic polarization curves obtained for HSLA steel after its immersion in 2.0 M H_2SO_4 solutions for (1) 10 min and (2) 60 min are shown in Figure 7. The corrosion potential (E_{corr}), corrosion current density (J_{corr}), cathodic ($_c$) and anodic ($_a$) Tafel slopes, polarization resistance (R_p), and corrosion rate (R_{corr}) that were obtained from polarization curves are listed in Table 2. The values of these parameters were obtained as previously reported [17–21]. It is clearly seen from Figure 7 and Table 2 that the increase of immersion time increases the values of J_{corr}, cathodic and anodic currents, and R_{corr}, while it decreases the values of R_p. This was due to the continuous dissolution of the HSLA steel under the harsh attack of the concentrated solution of the sulfuric acid, which does not allow the surface of steel to form protective layers and/or corrosion products. At this condition, the cathodic reaction for the steel in the sulfuric acid solution has been reported to be the hydrogen evolution reaction as follows [7, 9, 25]:

$$2H^+ + 2e^- = H_2$$

(2)

Table 2: Corrosion parameters obtained from the potentiodynamic polarization measurements for the HSLA steel electrode that was immersed for different periods of time in 2.0 M H_2SO_4 solutions with and without ATA molecules

Medium	Parameter						
	c	Ecorr	Jcorr	a	Rp	Rcorr	IE%
	mV dec−1	mV	µA cm−2	mV dec−1	Ωcm2	mm y−1	
2.0M H2SO4(10 min)	-90	-342	2750	65	3.73	27.57	—
+1.0 mM ATA (10 min)	-85	-335	1800	70	9.62	18.04	34.54

+5.0 mM ATA (10 min)	-82	-330	1300	75	19.4	13.03	52.73
2.0 M H2SO4(60 min)	-85	-325	4600	60	1.94	46.11	—
+1.0 mM ATA (60 min)	-78	-330	2400	63	5.93	24.06	47.83
+5.0 mM ATA (60 min)	-72	-325	1500	67	18.1	15.04	67.39

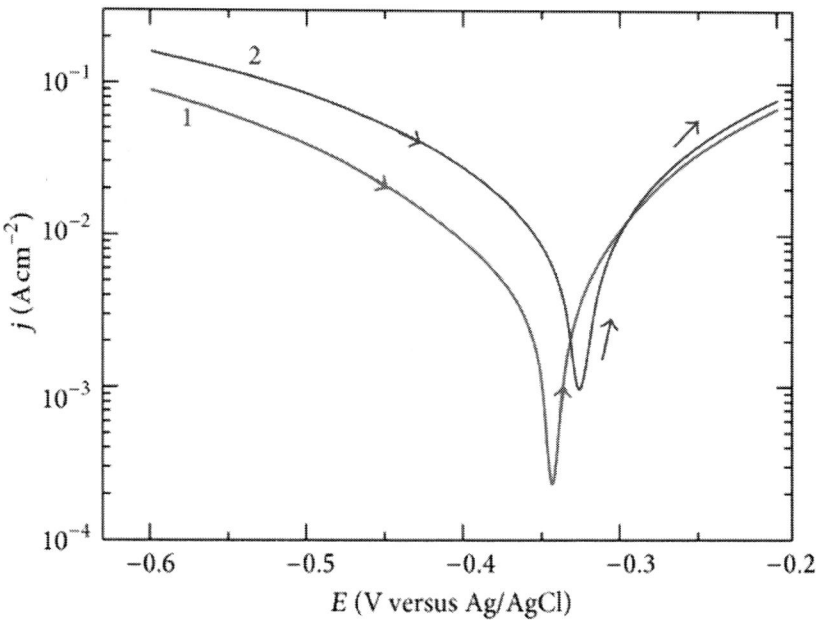

Figure 7: Potentiodynamic polarization curves obtained for the HSLA steel after its immersion for (1) 10 min and (2) 60 min in 2.0 M H_2SO_4 solutions.

On the other hand, the anodic reaction of the HSLA steel is the dissolution of its iron according to the following equation [7, 9]:

$$Fe = Fe^{2+} + 2e^{-}$$

(3)

The resulting ferrous cations (Fe^{2+}) are not stable and oxidize to ferric cations (Fe^{3+}) as follows:

$$Fe^{2+} = Fe^{3+} + e^-$$

(4)

The severity of these reactions increases with increase of the immersion time, which could lead to the increased dissolution of steel and also explain the increased currents and corrosion rate with increasing the time of immersion.

In order to evaluate the effect of ATA as a corrosion inhibitor after the different stated exposure intervals, the potentiodynamic polarization measurements were also carried out. Figure 8 shows the potentiodynamic polarization curves obtained for the HSLA steel after its immersion in 2.0 M H_2SO_4 solutions for 10 min in (1) the absence and the presence of (2) 1.0 mM and (3) 5.0 mM ATA, respectively. In order to study the effect of immersion time on the efficiency of ATA molecules, the polarization measurements were performed after 60 min and the curves are shown in Figure 9. The values of the parameters obtained from Figures 8 and 9 as well as the calculated values of IE% are also listed in Table 2. The values of IE% were obtained from the polarization data according to the following equation [9, 17, 18]:

$$IE\% = \frac{j_{Corr} - j^0_{Corr}}{j_{Corr}} \times 100,$$

(5)

where J_{corr} and j^0_{Corr} are the corrosion current densities in the absence and presence of ATA molecules, respectively.

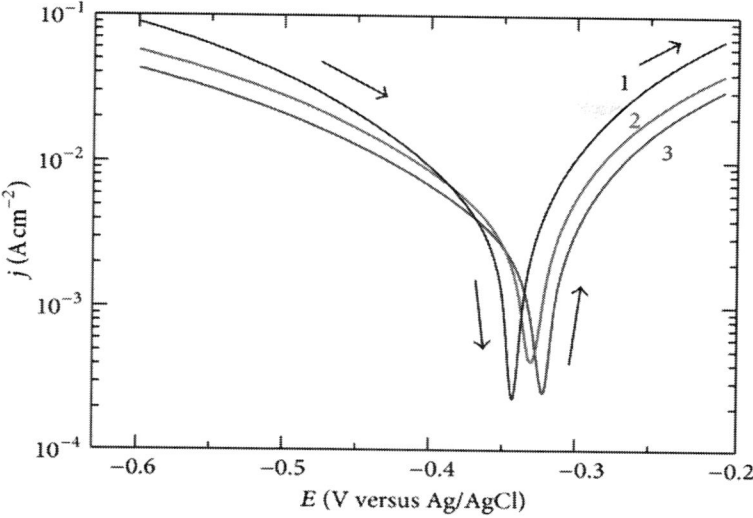

Figure 8: Potentiodynamic polarization curves obtained for the HSLA steel after its immersion for 10 min in 2.0 M H_2SO_4 in the absence (1) and the presence of (2) 1.0 mM ATA and (3) 5.0 mM ATA, respectively.

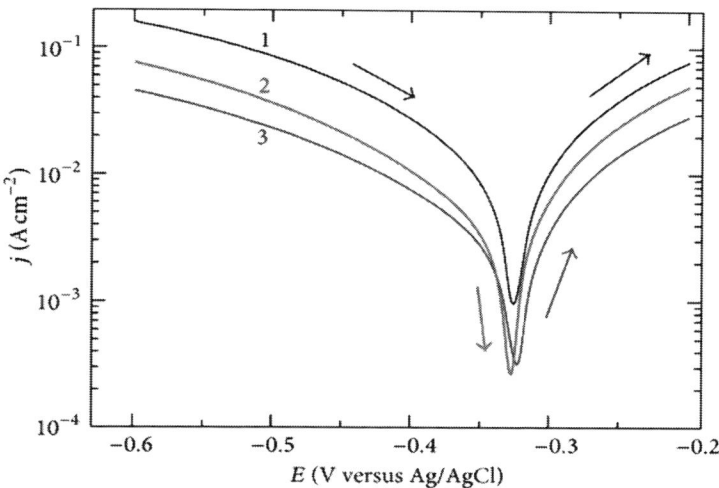

Figure 9: Potentiodynamic polarization curves obtained for the HSLA steel after its immersion for 60 min in 2.0 M H_2SO_4 in the absence (1) and the presence (2) of 1.0 mM ATA and (3) 5.0 mM ATA, respectively.

The addition of 1.0 mM ATA within the acid solution after all immersion periods of time remarkably decreased the anodic, cathodic, and corrosion currents. The data listed in Table 2 also indicated that the values of J_{Corr} and R_{Corr} decreased, while the value of R_p increased in the presence of 1.0 mM ATA compared to those recorded in its absence. This was perhaps due to the ability of ATA molecules to be adsorbed onto the steel surface, where the adsorption of ATA molecules onto the steel results in the formation of a protective layer that in turn not only isolates the surface but also blocks its active sites and thus precludes the corrosion of the steel in the corrosive 2.0 M sulfuric acid solution. It is also seen that the increase of ATA concentration to 5.0 mM greatly decreased the values of J_{Corr} and R_{Corr} and pronouncedly increased the value of R_p, particularly when the immersion time was increasing. This means that the increase of ATA concentration increases the adsorption probability of its molecules, which increases the efficiency of the formed layers in protecting the steel surface from being corroded easily. This was also indicated by the increase of the values of IE% with the increase of ATA concentration as listed in Table 2.

Although the corrosion of the HSLA steel increased with increasing the immersion time in 2.0 M H_2SO_4 solutions in the absence of ATA molecules, the corrosion of the steel was found to significantly decrease with the increase of immersion time in the presence of ATA and with the increase of its concentration. The increased corrosion of the HSLA steel in the absence of ATA was due to the rapid and harsh attack of the acid molecules toward the steel that makes its surface fresh, active, and dissolvable. On the other hand, the presence of ATA and the increase of its concentration strongly decrease the corrosion of the steel with increasing the immersion time as a result of decreasing the values of anodic and cathodic currents, J_{Corr} and R_{Corr}, and increasing the values of R_p. The decrease of the anodic and cathodic currents in the presence of ATA and with the increase of its concentration confirm that ATA is a mixed type corrosion inhibitor. The decrease of steel corrosion with time in the presence of ATA is due to the thickening of the adsorbed ATA layers onto the steel surface, which makes it more protected and precludes its dissolution. This was also confirmed by the large increase in IE% values with the increase of immersion time (see Table 2). The results obtained from the potentiodynamic polarization measurements therefore confirm those ones obtained by the EIS experiments and that the corrosion of HSLA steel increases with increasing the immersion time in the sulfuric acid solutions. It is also agreed that the addition of 1.0 mM ATA decreases the corrosion of steel and that effect increases with increasing both the concentration of ATA to 5.0 mM and the time of immersion from 10 min to 60 min.

CONCLUSIONS

The corrosion and corrosion control of HSLA steel in 2.0 M H_2SO_4 solutions using ATA as a corrosion inhibitor after different exposure intervals were reported. Electrochemical measurements indicated that the increase of immersion time from 10 min to 60 min increased the corrosion of the HSLA steel in the sulfuric acid solutions. On the other hand, the presence of ATA and the increase of its concentration were found to provide good corrosion inhibition and that effect increased with increasing the immersion time. This was confirmed by the increase of the polarization and solution resistance as well as the decrease of the anodic, cathodic, and corrosion currents, which in turn decreased the corrosion rate of HSLA in the acid medium. Moreover, the calculated value of the inhibition efficiency, IE%, was found to remarkably increase with increasing both ATA concentration and immersion time. Results collectively were in good agreement with each other showing clearly that the corrosion of HSLA steel increases with time and also ATA is a good mixed corrosion inhibitor due to the adsorption of its molecules onto the steel surface.

ACKNOWLEDGMENTS

The authors would like to extend their sincere appreciation to the Deanship of Scientific Research at King Saud University for its funding of this research through the Research Group Project no. RGP-VPP-160.

REFERENCES

1.   S. L. Chawla and R. K. Gupta, "Materials Selection for Corrosion Control," ASM International, 1993,http://www.asminternational. org/.

2. I. B. Timokhina, P. D. Hodgson, S. P. Ringer, R. K. Zheng, and E. V. Pereloma, "Precipitate characterisation of an advanced high-strength low-alloy (HSLA) steel using atom probe tomography,"Scripta Materialia, vol. 56, no. 7, pp. 601–604, 2007.

3. C. A. Melendres, N. Camillone III, and T. Tipton, "Laser raman spectroelectrochemical studies of anodic corrosion and film formation on iron in phosphate solutions," Electrochimica Acta, vol. 34, no. 2, pp. 281–286, 1989.

4. J. L. Yao, B. Ren, Z. F. Huang, P. G. Cao, R. A. Gu, and Z.-Q. Tian, "Extending surface Raman spectroscopy to transition metals for practical applications IV. A study on corrosion inhibition of benzotriazole on bare Fe electrodes," Electrochimica Acta, vol. 48, no. 9, pp. 1263–1271, 2003.

5. F. Bentiss, M. Traisnel, L. Gengembre, and M. Lagrenée, "Inhibition of acidic corrosion of mild steel by 3,5-diphenyl-4H-1,2,4-triazole," Applied Surface Science, vol. 161, no. 1, pp. 194–202, 2000.

6. F. B. Growcock and V. R. Lopp, "The inhibition of steel corrosion in hydrochloric acid with 3-phenyl-2-propyn-1-ol," Corrosion Science, vol. 28, no. 4, pp. 397–410, 1988.

7. E.-S. M. Sherif and A. H. Seikh, "Effects of immersion time and 5-Phenyl-1H-tetrazole on the corrosion and corrosion mitigation of cobalt free maraging steel in 0.5 M sulfuric acid pickling solutions," Journal of Chemistry, vol. 2013, Article ID 497823, 7 pages, 2013.

8. S. L. Granese, "Study of the inhibitory action of nitrogen-containing compounds," Corrosion, vol. 44, no. 6, pp. 322–327, 1988.

9. E.-S. M. Sherif, "Corrosion inhibition in 2.0 M sulfuric acid solutions of high strength maraging steel by aminophenyl tetrazole as a corrosion inhibitor," Applied Surface Science, vol. 292, pp. 190–196, 2014.

10. M. Lagrenée, B. Mernari, M. Bouanis, M. Traisnel, and F. Bentiss, "Study of the mechanism and inhibiting efficiency of 3,5-bis(4-methylthiophenyl)-4H-1,2,4-triazole on mild steel corrosion in acidic media," Corrosion Science, vol. 44, no. 3, pp. 573–588, 2002.

11. O. L. Riggs Jr., Corrosion Inhibitors, 2nd edition, edited by C. C. Nathan, National Association of Corrosion Engineers, Houston, Tex, USA, 1973.

12. M. Bartos and N. Hackerman, "A Study of inhibition action of propargyl alcohol during anodic dissolution of iron in hydrochloric acid," Journal of the Electrochemical Society, vol. 139, no. 12, pp. 3428–3433, 1992.

13. A. M. S. Abdennabi, A. I. Abdulhadi, S. T. Abu-Orabi, and H. Saricimen, "The inhibition action of 1(benzyl)1-H-4,5-dibenzoyl-1,2,3-triazole on mild steel in hydrochloric acid media," Corrosion Science, vol. 38, no. 10, pp. 1791–1800, 1996.

14. A. Chetouani, B. Hammouti, A. Aouniti, N. Benchat, and T. Benhadda, "New synthesised pyridazine derivatives as effective inhibitors for the corrosion of pure iron in HCl medium," Progress in Organic Coatings, vol. 45, no. 4, pp. 373–378, 2002.

15. M. Elayyachy, B. Hammouti, A. El Idrissi, and A. Aouniti, "Adsorption and corrosion inhibition behavior of C38 steel by one derivative of quinoxaline in 1 M HCl," Portugaliae Electrochimica Acta, vol. 29, no. 1, pp. 57–68, 2011.

16. A. Zarrouk, I. Warad, B. Hammouti, A. Dafali, S. S. Al-Deyab, and N. Benchat, "The effect of temperature on the corrosion of Cu/HNO$_3$ in the Presence of organic inhibitor: part-2," International Journal of Electrochemical Science, vol. 5, no. 10, pp. 1516–1526, 2010.

17. E.-S. M. Sherif, "Corrosion mitigation of copper in acidic chloride pickling solutions by 2-amino-5-ethyl-1,3,4-thiadiazole," Journal of Materials Engineering and Performance, vol. 19, no. 6, pp. 873–879, 2010.

18. E.-S. M. Sherif, "Comparative study on the inhibition of iron corrosion in aerated stagnant 3.5 wt % sodium chloride solutions by 5-phenyl-1H-tetrazole and 3-amino-1,2,4-triazole," Industrial and Engineering Chemistry Research, vol. 52, no. 41, pp. 14507–14513, 2013.

19. E.-S. M. Sherif, A. M. El Shamy, M. M. Ramla, and A. O. H. El Nazhawy, "5-(Phenyl)-4H-1,2,4-triazole-3-thiol as a corrosion inhibitor for copper in 3.5% NaCl solutions," Materials Chemistry and Physics, vol. 102, no. 2-3, pp. 231–239, 2007.

20. E.-S. M. Sherif and A. H. Ahmed, "Synthesizing new hydrazone derivatives and studying their effects on the inhibition of copper corrosion in sodium chloride solutions," Synthesis and Reactivity in Inorganic, Metal-Organic and Nano-Metal Chemistry, vol. 40, no. 6, pp. 365–372, 2010.

21. E. M. Sherif and S.-M. Park, "Inhibition of copper corrosion in acidic pickling solutions by N-phenyl-1,4-phenylenediamine," Electrochimica Acta, vol. 51, no. 22, pp. 4665–4673, 2006.

22. S. N. Banerjee and S. Misra, "1,10,-phenanthroline as corrosion inhibitor for mild steel in sulfuric acid solution," Corrosion, vol. 45, no. 9, pp. 780–783, 1989.

23. H. Ma, S. Chen, L. Niu, S. Zhao, S. Li, and D. Li, "Inhibition of copper corrosion by several Schiff bases in aerated halide solutions," Journal of Applied Electrochemistry, vol. 32, no. 1, pp. 65–72, 2002.

24. E.-S. M. Sherif, R. M. Erasmus, and J. D. Comins, "Corrosion of copper in aerated synthetic sea water solutions and its inhibition by 3-amino-1,2,4-triazole," Journal of Colloid and Interface Science, vol. 309, no. 2, pp. 470–477, 2007.

25. E. S. M. Sherif, "Corrosion behavior of duplex stainless steel alloy cathodically modified with minor ruthenium additions in concentrated sulfuric acid solutions," International Journal of Electrochemical Science, vol. 6, no. 7, pp. 2284–2298, 2011.

Corrosion Inhibition of Carbon Steel in HCl Solution by Some Plant Extracts

Ambrish Singh[1], Eno E. Ebenso[2], and M. A. Quraishi[1]

[1]Department of Applied Chemistry, Institute of Technology, Banaras Hindu University, Varanasi 221005, India

[2]Department of Chemistry, Faculty of Agriculture, Science & Technology, North West University (Mafikeng Campus), Mmabatho 2735, South Africa

ABSTRACT

The strict environmental legislations and increasing ecological awareness among scientists have led to the development of "green" alternatives to mitigate corrosion. In the present work, literature on green corrosion inhibitors has been reviewed, and the salient features of our work on green corrosion inhibitors have been highlighted. Among the studied leaves, extract Andrographis paniculata showed better inhibition performance (98%) than the other leaves extract. Strychnos nuxvomica showed better inhibition (98%) than the other seed extracts. Moringa oleifera is reflected as a good corrosion inhibitor of mild steel in 1 M HCl with 98% inhibition efficiency among the studied fruits extract. Bacopa monnieri showed its maximum inhibition performance to be 95% at 600 ppm among the investigated stem extracts. All the reported plant extracts were found to inhibit the corrosion of mild steel in acid media.

INTRODUCTION

Among the several methods of corrosion control and prevention, the use of corrosion inhibitors is very popular. Corrosion inhibitors are substances which when added in small concentrations to corrosive media decrease or prevent the reaction of the metal with the media. Inhibitors are added to many systems, namely, cooling systems, refinery units, chemicals, oil and gas production units, boiler, and so forth. Most of the effective inhibitors are used to contain heteroatom such as O, N, and S and multiple bonds in their molecules through which they are adsorbed on the metal surface. It has been observed that adsorption depends mainly on certain physicochemical properties of the inhibitor group, such as functional groups, electron density at the donor atom, π-orbital character, and the electronic structure of the molecule. Though many synthetic compounds showed good anticorrosive activity, most of them are highly toxic to both human beings and environment. The use of chemical inhibitors has been limited because of the environmental threat, recently, due to environmental regulations. These inhibitors may cause reversible (temporary) or irreversible (permanent) damage to organ system, namely, kidneys or liver, or disturbing a biochemical process or disturbing an enzyme system at some site in the body. The toxicity may be manifest either during the synthesis of the compound or during its applications. These known hazardous effects of most synthetic corrosion inhibitors are the motivation for the use of some natural products as corrosion inhibitors. Plant extracts have become important because they are environmentally acceptable, inexpensive, readily available and renewable sources of materials, and ecologically acceptable. Plant products are organic in nature, and some of the constituents including tannins, organic and amino acids, alkaloids, and pigments are known to exhibit inhibiting action. Moreover, they can be extracted by simple procedures with low cost. In the present work, the authors have reviewed literature on green corrosion inhibitors. Many authors such as E. E. Ebenso, B. Hammouti, A. Y. El Etre, P. C. Okafor, E. Oguzie, and P. B. Raja, have contributed significantly to the green mitigation by investigating several plants and their different body parts as corrosion inhibitors. The reviews of the literature along with salient features are summarised in Table 1.

Table 1: Plant extracts investigated as corrosion inhibitors by other authors

S. no.	Inhibitors used	Active constituents	Inhibition efficiency (%)	Remarks
(1)	Lawsonia		95.0	The aqueous extract of the leaves of henna (lawsonia) as the corrosion inhibitor was reported in C steel, nickel and zinc in acidic, neutral and alkaline solutions, using the polarization technique [1]
(2)	Fenugreek		92.2	The temperature effects were investigated on mild steel corrosion in 2.0 M of HCl and H_2SO_4 in the absence and presence of aqueous extract of fenugreek leaves (AEFLs) with the help of gravimetric method [2]
(3)	Olea europaea		93.0	The inhibitive action of the aqueous extract of olive leaves was reported towards the corrosion of C-steel in 2 M HCl solution using weight loss measurements, Tafel polarization, and cyclic voltammetry [3]
(4)	Cotula cinerea, Retama retam, and Artemisia herba	Anagyrine, cytisine	67.0	Plant extracts were investigated on the corrosion of X52 mild steel in aqueous 20% (2-3 M) sulphuric acid. Weight loss determinations and electrochemical measurements were also performed [4]
(5)	Eclipta alba		99.0	The inhibition effect of Eclipta alba in 1 N hydrochloric acid on corrosion of mild steel was investigated by weight loss, potentiodynamic polarization, and impedance methods, and the extracts of Eclipta alba were found to be effective corrosion pickling inhibitor [5]
(6)	Rauvolfia serpentina	Reserpine, ajmalicine, ajmaline, isoajmaline, ajmalinine, chandrine	94.0	Rauvolfia serpentina was tested as the corrosion inhibitor for mild steel in 1 M HCl and H_2SO_4 using weight loss method at three different temperatures namely, 303, 313, and 323 K. Potentiodynamic polarization, electrochemical impedance spectroscopy, and scanning electron microscope (SEM) studies were also performed [4]
(7)	Lupinus albus		86.5	The behaviour of the inhibitive effect of lupine (Lupinus albus L.) extract on the corrosion of steel in aqueous solution of 1 M sulphuric, and 2 M hydrochloric acid was studied by potentiodynamic polarization and electrochemical impedance spectroscopy (EIS) techniques [6]
(8)	Solanum tuberosum		91.3	The acid extracts of Solanum tuberosum were studied as the corrosion inhibitor for mild steel in 1 M HCl and H_2SO_4 medium using different techniques. It was found to be a good corrosion inhibitor [7]
(9)	Nauclea latifolia	Monoterpene, triterpene indole alkaloid, saponins	76.0	The inhibitive action of ethanol extracts from leaves (LV), bark (PK), and roots (RT) of Nauclea latifolia on mild steel corrosion in H_2SO_4 solutions at 30° and 60°C was studied using weight loss and gasometric techniques [8]
(10)	Sida rhombifolia		97.4	The efficacy of an acid extracts of leaves of Sida rhombifolia L. as the corrosion inhibitor for mild steel in 1 M phosphoric acid medium using weight loss measurements, polarization, and electrochemical impedance spectral studies were investigated. It was found to be an effective corrosion inhibitor [9]

(11)	*Ammi visnaga*		99.3	The inhibitive effect of the extract of Khillah (*Ammi visnaga*) seeds, on the corrosion of SX 316 steel in HCl solution using weight loss measurements as well as potentiostatic technique, was assessed. Negative values were calculated for the energy of adsorption indicating the spontaneity of the adsorption process [10]
(12)	*Embilica officianalis, Terminalia chebula* and *Terminalia bellirica*	Emblicanin A&B, puniglucanin, pedunculagin, tannic acid, chebulinic acid, and gallic acid	80%	Extracts were used in 5% (w:v) commercial hydrochloric acid as corrosion inhibitors of mild steel exposed into 5% (w:v) hydrochloric acid at 328 K on mild steel. Both Tafel polarization and linear polarization resistance techniques were used. Remarkable decrease in corrosion current and increase in linear polarization resistance values were observed in the presence of the acid extracts [11]
(13)	*Carica papaya* and *Acadirachta indica*	Papain, carpaine, chymopapain, azadirachtin, salannin, gedunin, and azadirone	87%	Extracts were used as corrosion inhibitors for corrosion of mild steel. The percentage inhibition of efficiency was found to increase with the increase in concentration of both inhibitors [12]
(14)	*Mentha pulegium*	Pulegone	80%	Natural oil extracted from pennyroyal mint (*Mentha pulegium*, PM) was evaluated as the corrosion inhibitor of steel in molar hydrochloric using weight loss measurements, electrochemical polarisation, and EIS methods. PM oil acted as an efficient cathodic inhibitor [13]
(15)	*Zanthoxylum alatum*	Terpineol, isoxazolidine, and imidazolinedione	85%	The inhibition effect of *Zanthoxylum alatum* plant extracts on the corrosion of mild steel in 5% and 15% aqueous hydrochloric acid solution was investigated by weight loss and electrochemical impedance spectroscopy (EIS) methods. The effect of temperature on the corrosion behaviour of mild steel in 5% and 15% HCl with the addition of plant extracts was studied in the temperature range 50–80°C. Surface analysis (SEM, XPS and FT-IR) was also carried out to establish the corrosion inhibitive property of this plant extract in HCl solution [14]

hyme, oriander, ibiscus, nis, Black umin and arden ress.	Thymol, malic acid, salicin, glutamic acid, leucine, and methionine	85%	Electrochemical impedance spectroscopy has been successfully used evaluate the performance of these compounds. The ac measurements that the dissolution process is activation controlled. Potentiodynami polarization curves indicate that the studied compounds are mixed-t inhibitors. Thyme, which contained the powerful antiseptic thymol a active ingredient, offers excellent protection for steel surface [15]
hoenix ctylifera, wsonia ermis, and ea mays	Lawsone, esculetin, fraxetin, allantoin, sterols, and hordenine	90%	Extracts were used as corrosion inhibitors for steel, aluminum, coppe brass in acid chloride and sodium hydroxide solutions using weight l solution analysis, and potential measurements. Only, *Phoenix dactyli Lawsonia inermis* extracts were found highly effective in reducing co rate of steel in acid chloride solutions and aluminum in sodium hydr solutions [16]
atura metel	Scopolamine, b-sitosterol, daturadiol, tropine, and daturilin	86%	Acid extract of the *D. metel* was studied for its corrosion inhibitive ef electrochemical and weight loss methods. The results of AC impedan polarisation studies correlate well with the weight loss studies [17]
icinus mmunis	Ricinoleic or ricinic acid, ricinolein, and palmitin	84%	The corrosion behaviour of plant extract (*Ricinus communis*) was stu means of electrochemical polarization, and impedance measurement of study from polarization and electrochemical impedance measuren indicated that *Ricinus communis* might alleviate the corrosion proces steel [18]
	Pugelone, alpha-pinene		*Mentha* was used as the corrosion inhibitor of steel in molar hydroch

rica *taya*	Chymopapain, pectin, carposide, carpaine, pseudocarpaine, dehydrocarpines, carotenoids, cryptoglavine, cis-violaxanthin, and antheraxanthin.	92%	Acid extracts of the different parts of *Carica papaya* were used as inh. various corrosion tests. Gravimetric and gasometric techniques were characterize the mechanism of inhibition [20]	
cia seyal	Catechu, dimethyltryptamine (DMT)	95%	The inhibitive effect of the gum exudate from *Acacia seyal* var. *seyal* studied on the corrosion of mild steel in drinking water using potentiodynamic polarization and electrochemical impedance spectr. (EIS) techniques. The corrosion rates of steel and inhibition efficiency gum exudates obtained from impedance and polarization measureme in good agreement [21]	
lotropis cera	a-and b-Amyrins, cyanidin-3-rhamnoglucoside, cycloart-23-en-3b, 25-diol, cyclosadol	89%	Extract of the *C. procera* was studied for its corrosion inhibitive effect weight loss, electrochemical, SEM, and UV methods. Using weight lo measurement data, mechanism of inhibitive action is probed by fittin adsorption isotherm [22]	
itella tica	Centellin, asiaticin, and centellicin	86%	*Centella asiatica* was studied as the corrosion inhibitor on mild steel i hydrochloric acid by weight loss method, gasometric method, potentiodynamic polarization method and AC impedance method [2	
ium ivum, lans regia l lostemon lin	Allyl cysteine sulfoxide, methyl allyl thiosulfinate, allicin, diallyl disulfide, diallyl trisulfide, ajoene, pogostone, friedelin, epifriedelinol, pachypodol, retusine, and oleanolic acid	94%	Plant extracts on the corrosion of steel in aqueous solution of I N sulp acid were studied by potentiodynamic polarization and electrochemi impedance spectroscopy (EIS) techniques [24]	
(27)	*Phyllanthus amarus*	Alkaloids, flavonoids, geraniin, hypophyllanthin, and phyllanthin	The inhibitive action of leaves (LV), seeds (SD), and a combination of leaves and seeds (LVSD) extracts of *Phyllanthus amarus* on mild steel corrosion in HCl and H$_2$SO$_4$ solutions was studied using weight loss and gasometric techniques. The results indicated that the extracts functioned as a good inhibitor in both environments and inhibition efficiency increased with extracts concentration. Temperature studies revealed an increase in inhibition efficiency with the rise in temperature, and activation energies decreased in the presence of the extract [26]	
(28)	*Azadirachta indica*	azadirachtin, azadirone, gedunin, nimbin, nimbandiol, nimbinene, nimbolide, nimonol, nimbolin, salannin,margolone, melianol, vilasanin, and flavanoids	80%	The inhibitive action of leaves (LV), root (RT), and seeds (SD) extracts of *Azadirachta indica* on mildsteel corrosion in H$_2$SO$_4$ solutions was studied using weight loss and gasometric techniques. The results obtained indicate that the extracts functioned as good inhibitors in H$_2$SO$_4$ solutions. Inhibition efficiency was found to increase with extracts concentration and temperature and followed the trend: SD > RT > LV. A mechanism of chemical adsorption of the phytochemical components of the plant extracts on the surface of the metal is proposed for the inhibition behaviour. The experimental data fitted into the Freundlich adsorption isotherm [27]
(29)	*Musa sapientum* and banana peels	Gallocatechin and dopamine	71%	The inhibition of the corrosion of mild steel by ethanol extract of *Musa sapientum* peels in H$_2$SO$_4$ was studied using gasometric and thermometric methods. The results of the study reveal that the different concentrations of ethanol extract of *M. sapientum* peels inhibit mild steel corrosion [28]
(30)	*Murraya koenigii*		80%	The inhibitive action of extract of curry leaves (*Murraya koenigii*) on carbon steel in 1N HCl was studied using weight loss, gasometric studies electrochemical polarization, and AC impedance measurements [29]
(31)	*Medicago Sativa*	biotin, cytidine, inosine, guanine, guanosine, and riboflavin	90%	The inhibitive effect of water and alcoholic extracts of *Medicago Sativa* (MS) on the corrosion of steel in 2.0 M H$_2$SO$_4$ containing 10% EtOH has been studied using chemical (weight loss (ML), hydrogen evolution (HE)), electrochemical (potentiodynamic polarization (PDP) and impedance spectroscopy (EIS)) techniques [30]

(32)	Oxandra asbeckii	Liriodenine, azafluorenones alkaloids	86%	The inhibition effect of alkaloids extract from Oxandra asbeckii plant (OAPE) on the corrosion of C38 steel in 1 M hydrochloric acid solution was investigated by potentiodynamic polarization and electrochemical impedance spectroscopy (EIS). The corrosion inhibition efficiency increases on increasing plant extracts concentration. Cathodic and anodic polarization curves showed that OAPE is a mixed-type inhibitor [31]
(33)	Adhatoda vasica, Eclipta alba, and Centella asiatica	Vasicine, vasicinone, asiaticoside, wedelolactone, β-sitosterol, and stigmasterol	99%	The inhibitive action of the extracts of Adhatoda vasica, Eclipta alba, and Centella asiatica on the corrosion of mild steel in 1N HCl was studied using weight loss method, electrochemical methods, and hydrogen permeation method. Polarization method indicated that the plant extracts are under mixed control, that is, promoting retardation of both anodic and cathodic reactions [32]
(34)	Ocimum sanctum, Aegle marmelos, and Solanum trilobatum		99%	A comparative study of the inhibitory effect of plant extracts, Ocimum sanctum, Aegle marmelos, and Solanum trilobatum, on the Corrosion of mild steel in 1N HCl medium was investigated using weight loss method, electrochemical methods, and hydrogen permeation method. Polarization method indicated that plant extracts behaved as mixed-type inhibitor [33]
(35)	Anna squamosa	Liriodenine and oxoanalobine	84%	Alkaloids extract from Annona squamosa plant has been studied as possible corrosion inhibitor for C38 steel in molar hydrochloric acid (1 M HCl). Potentiodynamic polarization and AC impedance methods have been used. The corrosion inhibition efficiency increases on increasing plant extract concentration [34]
(36)	Heinsia crinita			The paper provides information on the use of ethanol extract of Heinsia crinita as a corrosion inhibitor. Electrochemical studies such as polarisation and AC impedance spectra will throw more light on the mechanistic aspects of the corrosion inhibition [35]
(37)	Dacryodis edulis			The inhibition of low-carbon-steel corrosion in 1 M HCl and 0.5 M H_2SO_4 by extracts of Dacryodis edulis (DE) was investigated using gravimetric and electrochemical techniques. DE extract was found to inhibit the uniform and localized corrosion of carbon steel in the acidic media, affecting both the cathodic and anodic partial reactions [36]
(38)	Emblica officinalis		87%	Corrosion inhibition efficiency of acid extract of dry Emblica officinalis leaves for mild steel in 1N HCl medium was investigated. Experimental methods include weight loss, potentiodynamic polarization, and impedance studies [37]
(39)	Cyamopsis tetragonoloba	3-epikatonic acid 7-o-beta-(2-rhamnosyl-glucosyl) myricetin, ash, astragalin, caffeic acid, and chlorogenic acid	92%	The role of seed extract of Cyamopsis tetragonoloba on corrosion mitigation of mild steel in 1M HCl was investigated by weight loss method and potentiodynamic polarization technique. Experimental results were fitted into Langmuir and Temkin adsorption isotherm to study the process of inhibition [38]

In a previous work, the authors have investigated the extract of plants, namely, Azadirachta indica (leaves), Punica granatum (shell), and Momordica charantia as corrosion inhibitors on mild steel in 3% NaCl solution by chemical and electrochemical methods. Maximum inhibition efficiency of 86%, 82%, and 79% was obtained at a concentration of 6 mL/L, 3 mL/L and 1.2 mL/L, respectively. Azadirachta indica showed 97% antiscaling properties [39].

Aqueous extracts of Cordia latifolia and Curcumin were investigated as corrosion inhibitors for mild steel in industrial cooling systems. The extracts showed maximum inhibition efficiency of 97.7% and 60%, respectively [40].

The inhibitive effect of the aqueous extract of Jasmin (Jasminum auriculatum) on corrosion of mild steel in 3% NaCl was investigated. It showed inhibition efficiency of 80%. It was found to be predominantly the anodic corrosion inhibitor [41].

The inhibitive effects of aqueous extracts of Eucalyptus (leaves), Hibiscus (flower), and Agaricus on the corrosion of mild steel for cooling-water systems, using tap water, have been investigated by means of weight loss (under static as well as dynamic conditions) and polarization methods. All the plant extracts were found to inhibit corrosion of mild steel following and their inhibitive efficiencies were in the order: Agaricus (85%), Hibiscus (79%), and Eucalyptus (74%) under the static test conditions. The inhibition efficiencies remain almost the same under the dynamic test conditions, which are nearer to field conditions. All the inhibitors (extracts) were found to follow Langmuir as well as Freundlich adsorption isotherms, that is, they inhibit corrosion through adsorption. Polarization measurements gave a similar order of inhibition efficiencies of plant extracts as that determined using the weight loss technique. Agaricus extract was found to be predominantly a cathodic inhibitor, while the extracts of Eucalyptus and Hibiscus were found to be mixed inhibitors [40].

Ascorbic acid in combination with DQ-2000 (aminotrimethyl phosphonic acid) and DQ-2010 (1-hydroxyethylidine 1,1-diphosphonic acid) was used to reduce the concentration of zinc in the blowdown of the cooling systems. All the inhibitors used were found to be effective. The maximum inhibition efficiency 99.2% was obtained with DQ-2010 100 ppm + Ascorbic acid 200 ppm concentration. Inhibitors follow Langmuir isotherm which showed that they inhibit corrosion through adsorption [42].

In present work, authors have used the extract of (Kalmegh) Andrographis paniculata, (Meethi Neem)Murraya koenigii, (Bael) Aegle marmelos, (Kuchla) Strychnos nuxvomica, (Karanj) Pongamia pinnata, (Jamun)Syzygium cumini, (Shahjan) Moringa oleifera, (Pipali) Piper longum, (Orange) Citrus aurantium, (Brahmi)Bacopa monnieri, (Pipal) Ficus religiosa, and (Arjun) Terminalia arjuna as corrosion inhibitors [43–48]. The active constituents and inhibition efficiencies of the extracts used are summarized in Table 2.

Table 2: Plant extracts used by us as corrosion inhibitors

S. no.	Plant used	Active constituents	Common name	Inhibition efficiency (%)
(A)	*Murraya koenigii*			96.7
(1)			Murrafoline-I	
(2)			Pyrayafoline-D	
(3)			Mahabinine-A	
(B)	*Aegle marmelos*			96.2
(1)			Skimmianine	
(C)	*Andrographis paniculata*			98.1
(1)			Andrographolide	
(D)	*Syzygium cumini*			94.2
(1)			Ellagic acid	
(2)			Gallic acid	
(3)			Quercetin	
(4)			Cafeic acid	

| (E) | *Ponganıia pinnata* | | 97.6 |

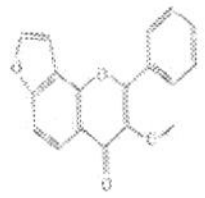

(1) Karanjin

(2) Pongapine

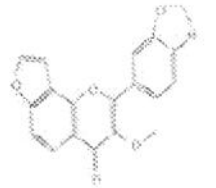

(3) Kanjone

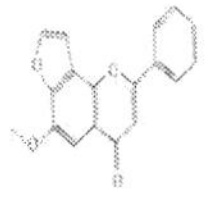

(4) Pongaglabrone

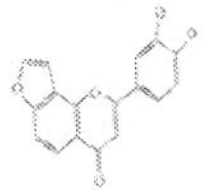

| (F) | *Strychnos nuxvomica* | 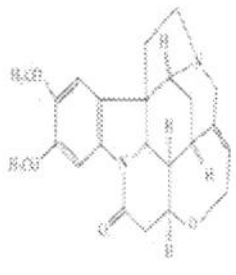 Brucine | 98.2 |

| (G) | *Piper longum* | | 97.6 |

(1) Piperine

(2) Piplartine

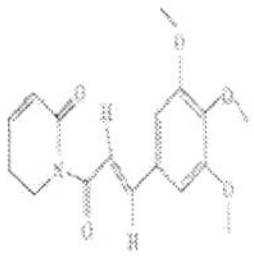

(3) Rutin

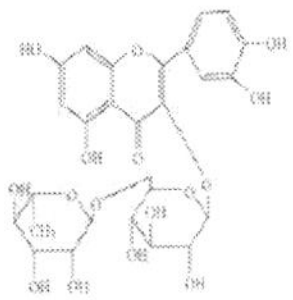

| (H) | *Moringa oleifera* | | 98.6 |

(1) Arginine

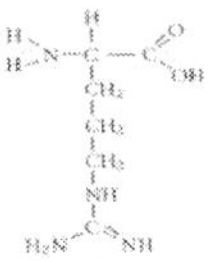

| (I) | *Citrus Aurantium* | | 89.6 |

(1) Threonine

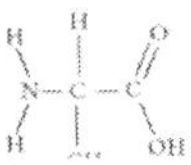

(J)	*Terminalia arjuna*		88.9
(1)		b-Sitosterol	
(K)	*Ficus religiosa*		88.8
(1)		Lanosterol	
(L)	*Bacopa monnieri*		95.2
(1)		Bacoside A	
(2)		Bacoside B	

EXPERIMENTAL

Prior to all measurements, the mild steel specimens, having composition (in wt%) 0.076 C, 0.012 P, 0.026 Si, 0.192 Mn, 0.050 Cr, 0.135 Cu, 0.023 Al, 0.050 Ni, and the remainder iron, were polished successively with fine grade Emery papers from 600 to 1200 grades. The specimens were washed thoroughly with double-distilled water and finally degreased with acetone and dried at room temperature. The aggressive solution 1 M HCl was prepared by dilution of analytical grade HCl (37%) with double-distilled water, and all experiments were carried out in unstirred solutions.

AC impedance (EIS) measurements and potentiodynamic polarization studies were carried out using a GAMRY PCI 4/300 electrochemical work station based on ESA 400. Gamry applications include EIS 300 (for EIS measurements) and DC 105 software (for corrosion) and Echem Analyst (5.50 V) software for data fitting. All electrochemical experiments were performed in a Gamry three-electrodes electrochemical cell under the atmospheric conditions with a platinum counter electrode and a saturated calomel electrode (SCE) as the reference electrode. The working electrode mild steel (7.5 cm long stem) with the exposed surface of $1.0 \, cm^2$ was immersed into aggressive solutions with and without inhibitor, and then the open circuit potential was measured after 30 minutes. EIS measurements were performed at corrosion potentials, E_{corr}, over a frequency range of 100 kHz to 10 mHz with an AC signal amplitude perturbation of 10 mV peak to peak. Potentiodynamic polarization studies were performed with a scan rate of 1 mVs^{-1} in the potential range from 250 mV below the corrosion potential to 250 mV above the corrosion potential. All potentials were recorded with respect to the SCE.

RESULTS AND DISCUSSION

Leaves Extract as Corrosion Inhibitors

The leaves extract of Andrographis paniculata, Murraya koenigii, and Aegle marmelos were investigated as corrosion inhibitors by weight loss and electrochemical methods in the present study. Among the studied leaves extract, Andrographis paniculata showed better inhibition performance than the other leaves extract. The result is summarized in Table 3 and Figure 1. The order of their inhibition efficiency has been found as follows:

Andrographis paniculata

> Murraya koenigii > Aegle marmelos.

(1)

Table 3: Electrochemical impedance and Tafel data at 308 K

Name of inhibitor	Inhibitor concentration	R_{ct} (Ω cm^2)	C_{dl}(μF cm^{-2})	E(%)	-E_{corr} (mV versus SCE)	i_{corr}(mA/ cm^2)	E(%)
1 M HCl	—	8.5	68.9	—	446	1540.0	—

Murraya koenigii	240.0	180.3	59.0	95.3	480	71.0	95.5
	300.0	256.2	58.2	96.6	469	48.0	96.9
	600.0	344.3	50.5	97.5	472	47.0	97.0
Aegle marmelos	200.0	101.9	59.2	91.7	457	159.0	89.3
	300.0	151.1	44.1	94.4	466	100.0	93.5
	400.0	264.8	30.7	96.7	499	60.0	96.0
Andrographis paniculata	300.0	99.0	56.9	91.4	489	82.0	94.6
	600.0	108.0	52.4	92.1	462	59.0	96.1
	1200.0	491.0	40.4	98.2	486	30.6	98.0

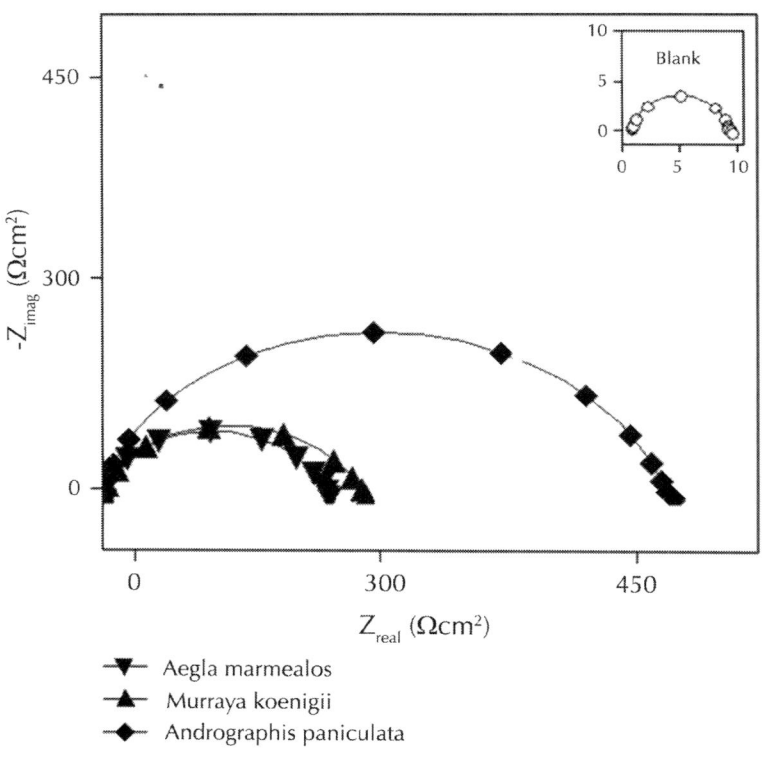

(a)

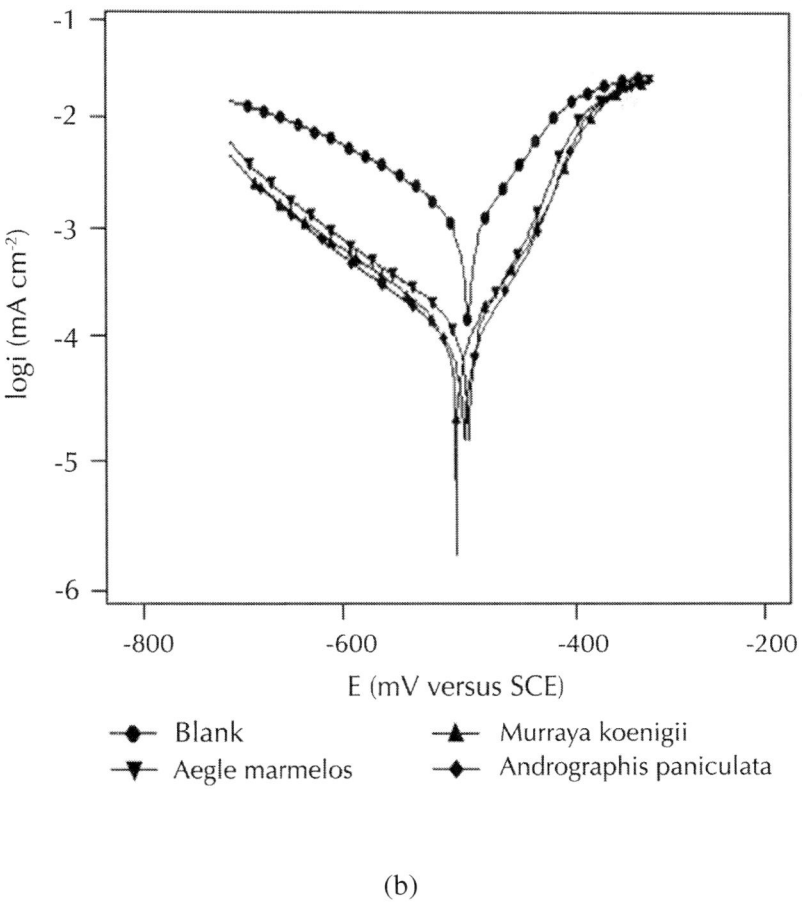

(b)

Figure 1: Nyquist plots and Tafel plots for mild steel in 1 M HCl in the absence and presence of different inhibitors at their optimum concentration.

The higher inhibitive performance of Andrographis paniculata is due to the presence of delocalized π-electrons. This extensive delocalized π-electrons favours its greater adsorption on the mild steel surface, thereby giving rise in very high inhibition efficiency (98.1%) at a concentration of 1200 ppm the relatively better performance of Murraya koenigii (96.7%) at 600 ppm than Aegle marmelos (96.2%) at 400 ppm. The most pronounced effect and the highest R_{ct} value (491.0 ohm cm²) was obtained by inhibitor Andrographis paniculata at 1200 ppm concentration. The lowest R_{ct} value (264.8 ohm cm²) was obtained by inhibitor Aegle marmelos. The high R_{ct} values are generally associated with a slower corroding system. These data revealed that R_{ct} values increased after the addition of inhibitors, and on the other hand, C_{dl} values decreased. This situation was a result of the adsorption of inhibitors at the metal/solution interface. A decrease in local dielectric constant and/or an increase in the thickness of the electrical double layer can cause this decrease in C_{dl} values, suggesting that the water molecules (having high dielectric constant) are replaced with inhibitor molecules (having low dielectric constant). It is worth noting that the percentage inhibition efficiencies obtained from impedance measurements were reasonably in a good agreement with those obtained from weight loss measurements.

Seed Extracts as Corrosion Inhibitors

We have used seed extracts of Strychnos nuxvomica, Pongamia pinnata, and Syzygium cumini in our present study. The result is concluded in Table 4 and Figure 2. The order of their inhibition efficiency has been found as follows:

$$Strychnos\ nuxvomica > Pongamia\ pinnata > Syzygium\ cumini. \tag{2}$$

Table 4: Electrochemical impedance, Tafel, and linear polarization resistance data at 308 K

Name of inhibitor	Inhibitor concentration	R_{ct} ($\Omega\,cm^2$)	C_{dl} ($\mu F\,cm^{-2}$)	E (%)	$-E_{corr}$ (mV versus SCE)	i_{corr} (mA/ cm^2)	E (%)
1 M HCl	—	8.5	68.9	—	446	1540.0	—
Syzygium cumini	240.0	97.1	67.6	91.2	443	165.0	89.2
	300.0	117.5	56.1	92.7	462	98.0	93.5
	600.0	238.5	53.7	96.4	469	60.0	96.0
Pongamia pinnata	300.0	129.5	39.6	92.9	461	84.0	94.0
	350.0	150.6	38.7	93.5	482	77.0	95.0
	400.0	239.8	35.7	96.5	471	49.0	97.0
Strychnos nuxvomica	250.0	130.3	52.0	93.5	461	132.0	91.4
	300.0	159.9	47.1	94.7	463	97.0	93.7
	350.0	263.9	43.3	96.7	494	27.5	98.2

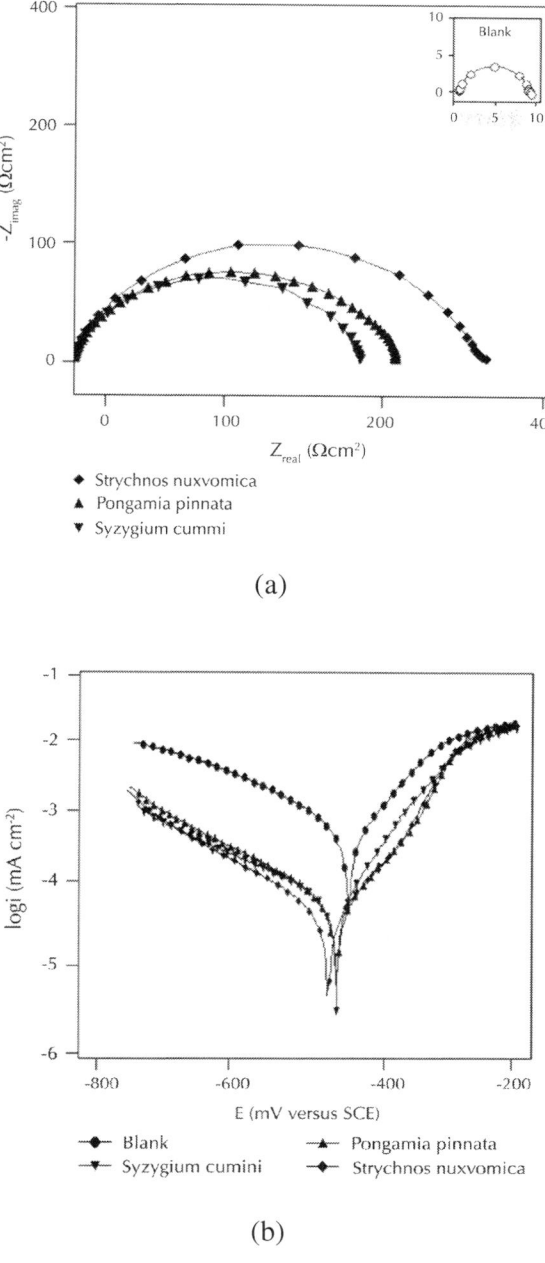

Figure 2: Nyquist plots and Tafel plots for mild steel in 1 M HCl in the absence and presence of different inhibitors at their optimum concentrations.

The best performance of Strychnos nuxvomica as the corrosion inhibitor can be attributed to the presence of three methoxy groups attached to the benzene nucleus. These extensive groups favor its greater adsorption on the mild steel surface, thereby giving rise to very high inhibition efficiency (98.2%) at a concentration as low as 350 ppm. The next best performance of Pongamia pinnata (97.6%) has been found at 400 ppm concentration. It was found that R_{ct} values increased to a maximum of 264 (Ω cm^2) at an optimum concentration of Strychnos nuxvomica. This situation was a result of the adsorption of inhibitors at the metal/solution interface. In the present study, maximum displacement was 48 mV, suggesting that tested seeds extract belonged to the mixed-type inhibitors.

Fruits Extracts as Corrosion Inhibitors

We have used fruits extract of Moringa oleifera, Piper longum and Citrus aurantium in our present study. The result is depicted in Table 5 and Figure 3. The inhibition efficiency of fruits extract follows the order

Moringa oleifera

> Piper longum > Citrus aurantium

(3)

Table 5: Electrochemical impedance, Tafel, and linear polarization resistance data at 308 K

Name of inhibitor	Inhibitor concentration	R_{ct} (Ω cm^2)	C_{dl} (μF cm^{-2})	E (%)	$-E_{corr}$ (mV versus SCE)	i_{corr} (mA/ cm^2)	E (%)
1 M HCl	—	8.5	68.9	—	446	1540.0	—
Piper longum	240.0	213.2.1	46.4	96.0	464	53.0	96.5
	300.0	273.3	33.1	96.9	469	46.0	96.9
	600.0	355.5	27.3	97.6	479	41.0	97.3
Moringa oleifera	200.0	215.0	43.0	96.0	503	59.0	96.1
	250.0	324.5	41.4	97.3	472	38.0	97.5
	300.0	644.9	32.4	98.6	493	28.0	98.1
Citrus aurantium	300.0	23.5	68.5	68.9	466	430.0	72.0
	600.0	58.2	65.4	85.4	515	212.0	86.2
	1200.0	65.2	56.3	87.0	464	160.0	89.6

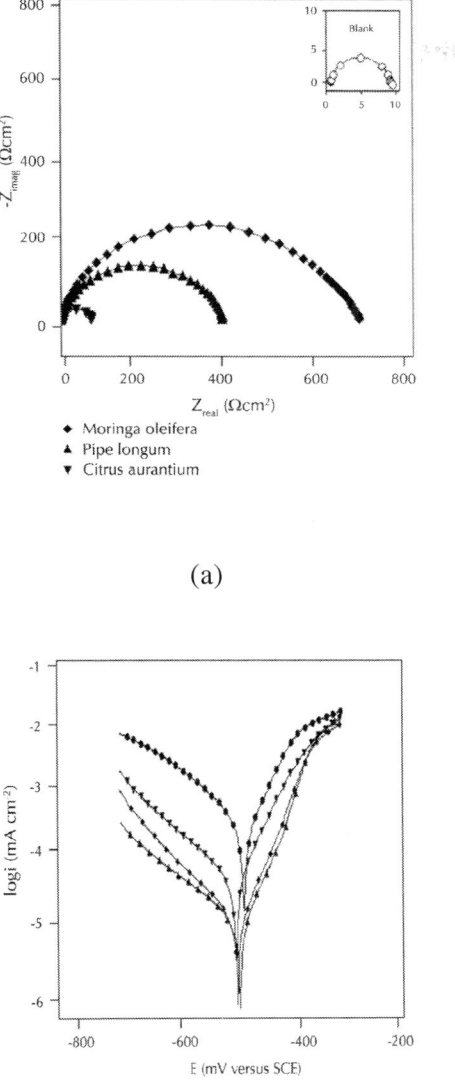

(a)

(b)

Figure 3: Nyquist plots and Tafel plots for mild steel in 1 M HCl in the absence and presence of different inhibitors at their optimum concentrations.

Good performance of fruits extract as corrosion inhibitors for mild steel in 1 M HCl solutions may be due to the presence of heteroatoms, π-electrons, and aromatic rings in their structures. The highest inhibition efficiency shown by Moringa oleifera is 98.2% at 300 ppm due to the presence of imine (C=N) group, four N atoms, and long alkyl chain and least efficiency of Citrus aurantium is 88.1% at 1200 ppm attributed to the presence of electron withdrawing COOH group. The R_{ct} values were found to increase, and on the other hand, C_{dl} values decreased in the presence of all fruits extract. This is due to the adsorption of these compounds at the metal/solution interface. The values of I_{corr} were found to decrease in the presence of inhibitors. The decrease in I_{corr} values can be due to the adsorption of fruits extract on the mild steel surface. It was observed that there is a small shift towards the cathodic region in the values of E_{corr}. In the present study, maximum displacement in E_{corr} value was 69 mV, which indicates that all studied fruits extract were mixed-type inhibitors.

Stem Extracts as Corrosion Inhibitors

Stem extracts of Bacopa monnieri, Ficus religiosa, and Terminalia arjuna were used as corrosion inhibitors.Bacopa monnieri showed its maximum inhibition performance 95.2% at 600 ppm, while Ficus religiosa shows 88.7% at 1200 ppm. The better performance of Bacopa monnieri can be attributed to the presence of more O atoms in its structure. Terminalia arjuna has been found to give its maximum inhibition efficiency 83.4% at 1200 ppm. The R_{ct} values were found to increase and on the other hand, C_{dl} values decreased in the presence of all stem extract as in Table 6 and Figure 4. This may be due to the adsorption of these compounds at the metal/solution interface. Decrease in C_{dl} values, caused by a decrease in local dielectric constant and/or an increase in the thickness of the electrical double layer, suggests that the water molecules are replaced by inhibitor molecules. It was observed that the values of I_{corr} decrease in the presence of inhibitors. The decrease in I_{corr} values can be due to the adsorption of stems extract on the mild steel surface. The and values remained more or less identical in the absence and presence of stems extract studied, suggesting that the effect of inhibitors is not as large as to change the mechanism of corrosion.

Table 6: Electrochemical impedance, Tafel, and linear polarization resistance data at 308 K

Name of inhibitor	Inhibitor concentration	R_{ct} ($\Omega\,cm^2$)	C_{dl} ($\mu F\,cm^{-2}$)	E (%)	$-E_{corr}$ (mV versus SCE)	i_{corr} (mA/ cm^2)	E (%)
1 M HCl	—	8.5	68.9	—	446	1540.0	—
Terminalia arjuna	300.0	17.0	67.4	50.5	478	785.0	49.0
	600.0	26.2	48.9	67.9	461	713.0	53.7
	1200.0	75.9	38.8	88.9	469	220.0	85.7
Ficus religiosa	300.0	28.7	63.9	70.7	444	407.0	54.0
	600.0	37.8	63.0	77.7	481	301.0	80.4
	1200.0	75.6	37.6	88.8	464	190.0	87.6
Bacopa monnieri	240.0	41.9	53.5	79.9	464	518.0	66.3
	300.0	74.2	44.2	88.6	486	218.0	85.8
	600.0	175.2	39.4	95.2	489	75.0	95.1

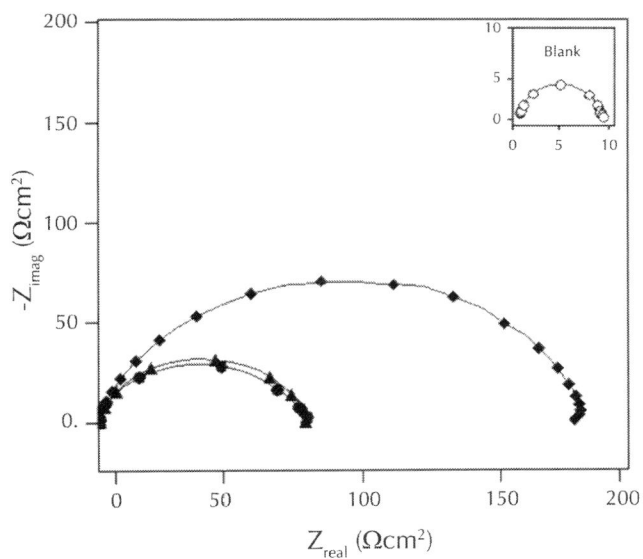

♦ Bacopa momieri
▲ Ficus religiosa
● Terminalia arjuna

(a)

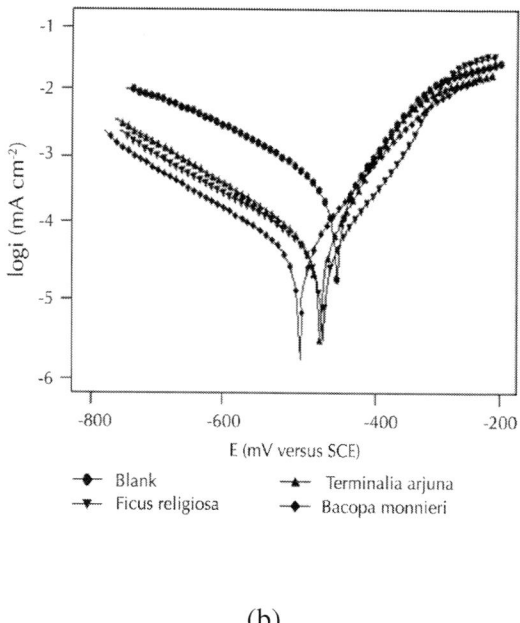

(b)

Figure 4: Nyquist plots and Tafel plots for mild steel in 1 M HCl in the absence and presence of different inhibitors at their optimum concentrations.

All the studied plant extracts obtained from leaves, seeds, fruits, and stem showed good inhibition efficiency (>95%) at their optimum concentrations for mild steel in 1 M HCl. The optimum concentration is considered as a concentration beyond which increase in extract concentration showed no significant change in the inhibition efficiency. The good performance may be attributed to the synergism between the different compounds present in the extracts. Andrographis paniculata leaves extract showed 98% inhibition efficiency due to the presence of delocalized π-electrons as compared to those of Strychnous nuxvomica seed extract which can be attributed to the presence of three methoxy groups attached to the benzene nucleus favoring its greater adsorption on the mild steel surface, thereby giving rise to very high inhibition efficiency (98.2%) andMoringa oleifera fruit extract (98.1%) due to the presence of imine (C=N) group, four N atoms and long alkyl chain. Also, the low inhibition efficiency of Bacopa monnieri as compared to Andrographis paniculata,Strychnous nuxvomica, and Moringa oleifera can be attributed to the presence of O atoms in its structure.

Mechanism of Corrosion Inhibition

In acidic solutions, transition of the metal/solution interface is attributed to the adsorption of the inhibitor molecules at the metal/solution interface, forming a protective film. The rate of adsorption is usually rapid, and hence, the reactive metal surface is shielded from the acid solutions [49]. The adsorption of an inhibitor depends on its chemical structure, its molecular size, the nature and charged surface of the metal, and distribution of charge over the whole inhibitor molecule. In fact, adsorption process can occur through the replacement of solvent molecules from the metal surface by ions and molecules accumulated near the metal/solution interface. Ions can accumulate at the metal/solution interface in excess of those required to balance the charge on the metal at the operating potential. These ions replace solvent molecules from the metal surface, and their centres reside at the inner Helmholtz plane. This phenomenon is termed specific adsorption, contact adsorption. The anions are adsorbed when the metal surface has an excess positive charge in an amount greater than that required to balance the charge corresponding to the applied potential. The exact nature of the interactions between a metal surface and an aromatic molecule depends on the relative coordinating strength towards the given metal of the particular groups present [50].

Generally, two modes of adsorption were considered. In one mode, the neutral molecules of leaves extract can be adsorbed on the surface of mild steel through the chemisorption mechanism, involving the displacement of water molecules from the mild steel surface and the sharing electrons between the heteroatoms and iron. The inhibitor molecules can also adsorb on the mild steel surface based on donor-acceptor interactions between π-electrons of the aromatic/heterocyclic ring and vacant d-orbitals of surface iron. In another mode, since it is well known that the steel surface bears the positive charge in acidic solutions [51], so it is difficult for the protonated leaves extract to approach the positively charged mild steel surface (H_3O^+/metal interface) due to the electrostatic repulsion. Since chloride ions have a smaller degree of hydration, thus they could bring excess negative charges in the vicinity of the interface and favour more adsorption of the positively charged inhibitor molecules, the protonated leaves extract adsorbed through electrostatic interactions between the positively charged molecules and the negatively charged metal surface.

Since all the different parts of plant extract possess several heteroatoms containing active constituents, therefore there may be a synergism between the molecules accounting for the good inhibition efficiencies.

CONCLUSIONS

- All the extracts studied showed good inhibition efficiency.
- Andrographis paniculata, Strychnous nuxvomica, and Moringa oleifera extracts showed inhibition efficiency above 98%.
- All the extracts were found to be the mixed type of inhibitors.
- All the results obtained from EIS, LPR, and weight loss are in good agreement with each other.

REFERENCES

1.   A. Y. El-Etre, M. Abdallah, and Z. E. El-Tantawy, "Corrosion inhibition of some metals using lawsonia extract," Corrosion Science, vol. 47, no. 2, pp. 385–395, 2005. ·

2.   E. A. Noor, "Temperature effects on the corrosion inhibition of mild steel in acidic solutions by aqueous extract of fenugreek leaves," International Journal of Electrochemcal Science, vol. 2, pp. 996–1017, 2007.

3.   A. Y. El-Etre, "Inhibition of acid corrosion of carbon steel using aqueous extract of olive leaves," Journal of Colloid and Interface Science, vol. 314, no. 2, pp. 578–583, 2007. · ·

4.   P. B. Raja and M. G. Sethuraman, "Natural products as corrosion inhibitor for metals in corrosive media—A review," Materials Letters, vol. 62, no. 1, pp. 113–116, 2008. ·

5.   M. Shyamala and A. Arulanantham, "Eclipta alba as corrosion pickling inhibitor on mild steel in hydrochloric acid," Journal of Materials Science and Technology, vol. 25, no. 5, pp. 633–636, 2009.

6.   A. M. Abdel-Gaber, B. A. Abd-El-Nabey, and M. Saadawy, "The role of acid anion on the inhibition of the acidic corrosion of

steel by lupine extract," Corrosion Science, vol. 51, no. 5, pp. 1038–1042, 2009. ·

7.   P. Bothi Raja and M. G. Sethuraman, "Solanum tuberosum as an inhibitor of mild steel corrosion in acid media," Iranian Journal of Chemistry and Chemical Engineering, vol. 28, no. 1, pp. 77–84, 2009.

8.   I. E. Uwah, P. C. Okafor, and V. E. Ebiekpe, "Inhibitive action of ethanol extracts from Nauclea latifoliaon the corrosion of mild steel in H_2SO_4 solutions and their adsorption characteristics," Arabian Journal of Chemistry. In press. ·

9.   R. Saratha and R. Meenakshi, "Corrosion inhibitor-A plant extract," Der Pharma Chemica, vol. 2, pp. 287–294, 2010.

10. A. Y. El-Etre, "Khillah extract as inhibitor for acid corrosion of SX 316 steel," Applied Surface Science, vol. 252, no. 24, pp. 8521–8525, 2006. ·

11. M. J. Sanghvi, S. K. Shukla, A. N. Misra, M. R. Padh, and G. N. Mehta, "Inhibition of hydrochloric acid corrosion of mild steel by aid extracts of embilica officianalis, terminalia bellirica and terminalia chebula," Bulletin of Electrochemistry, vol. 13, no. 8-9, pp. 358–361, 1997.

12. E. E. Ebenso, J. Udofot, J. Ekpe, and U. J. Ibok, "Studies on the inhibition of mild steel corrosion by some plant extracts in acidic medium," Discovery and Innovation, vol. 10, no. 1-2, pp. 52–59, 1998.

13. A. Bouyanzer, B. Hammouti, and L. Majidi, "Pennyroyal oil from Mentha pulegium as corrosion inhibitor for steel in 1 M HCl," Materials Letters, vol. 60, no. 23, pp. 2840–2843, 2006. ·

14. L. R. Chauhan and G. Gunasekaran, "Corrosion inhibition of mild steel by plant extract in dilute HCl medium," Corrosion Science, vol. 49, no. 3, pp. 1143–1161, 2007. ·

15. E. Khamis and N. Alandis, "Herbs as new type of green inhibitors for acidic corrosion of steel,"Materialwissenschaft und Werkstofftechnik, vol. 33, no. 9, pp. 550–554, 2002. ·

16. H. H. Rehan, "Corrosion control by water-soluble extracts from leaves of economic plants,"Materialwissenschaft und Werkstofftechnik, vol. 34, no. 2, pp. 232–237, 2003. ·

17. M. G. Sethuraman and P. B. Raja, "Corrosion inhibition of mild steel by Datura metel in acidic medium," Pigment and Resin Technology, vol. 34, no. 6, pp. 327–331, 2005. ·

18. R. A. L. Sathiyanathan, M. M. Essa, S. Maruthamuthu, M. Selvanayagam, and N. Palaniswamy, "Inhibitory effect of Ricinus communis (Castor-oil plant) leaf extract on corrosion of mild steel in low chloride medium," Journal of the Indian Chemical Society, vol. 82, no. 4, pp. 357–359, 2005.

19. E. Chaieb, A. Bouyanzer, B. Hammouti, and M. Benkaddour, "Inhibition of the corrosion of steel in 1 M HCl by eugenol derivatives," Applied Surface Science, vol. 246, no. 1–3, pp. 199–206, 2005. ·

20. P. C. Okafor and E. E. Ebenso, "Inhibitive action of Carica papaya extracts on the corrosion of mild steel in acidic media and their adsorption characteristics," Pigment and Resin Technology, vol. 36, no. 3, pp. 134–140, 2007. ·

21. J. Buchweishaija and G. S. Mhinzi, "Natural products as a source of environmentally friendly corrosion inhibitors: the case of gum exudate from Acacia seyal var. seyal," Portugaliae Electrochimica Acta, vol. 26, no. 3, pp. 257–265, 2008.

22. P. B. Raja and M. G. Sethuraman, "Inhibition of corrosion of mild steel in sulphuric acid medium byCalotropis procera," Pigment and Resin Technology, vol. 38, no. 1, pp. 33–37, 2009. ·

23. M. Shyamala and A. Arulanantham, "Corrosion inhibition effect of centella asiatica (Vallarai) on mild steel in hydrochloric acid," Asian Journal of Chemistry, vol. 21, no. 8, pp. 6102–6110, 2009.

24. C. Anca, M. Ioana, D. I. Vaireanu, L. Iosif, L. Carmen, and C. Simona, "Estimation of inhibition efficiency for carbon steel corrosion in acid media by using natural plant extracts," Revista de Chimie, vol. 60, no. 11, pp. 1175–1180, 2009.

25. P. C. Okafor, I. E. Uwah, O. O. Ekerenam, and U. J. Ekpe, "Combretum bracteosum extracts as eco-friendly corrosion inhibitor for mild steel in acidic medium," Pigment and Resin Technology, vol. 38, no. 4, pp. 236–241, 2009. ·

26. P. C. Okafor, M. E. Ikpi, I. E. Uwah, E. E. Ebenso, U. J. Ekpe, and S. A. Umoren, "Inhibitory action ofPhyllanthus amarus extracts on the corrosion of mild steel in acidic media," Corrosion Science, vol. 50, no. 8, pp. 2310–2317, 2008. ·

27. P. C. Okafor, E. E. Ebenso, and U. J. Ekpe, "Azadirachta indica extracts as corrosion inhibitor for mild steel in acid medium," International Journal of Electrochemical Science, vol. 5, no. 7, pp. 978–993, 2010.

28. N. O. Eddy and E. E. Ebenso, "Adsorption and inhibitive properties of ethanol extracts of Musa sapientum peels as a green corrosion inhibitor for mild steel in H_2SO_4," African Journal of Pure and Applied Chemistry, vol. 2, pp. 046–054, 2008.

29. A. Sharmila, A. A. Prema, and P. A. Sahayaraj, "Influence of Murraya koenigii (curry leaves) extract on the corrosion inhibition of carbon steel in HCL solution," Rasayan Journal of Chemistry, vol. 3, no. 1, pp. 74–81, 2010.

30. A. M. Al-Turkustani, S. T. Arab, and L. S. S. Al-Qarni, "Medicago Sative plant as safe inhibitor on the corrosion of steel in 2.0 M H_2SO_4 solution," Journal of Saudi Chemical Society, vol. 15, no. 1, pp. 73–82, 2011. ·

31. M. Lebrini, F. Robert, A. Lecante, and C. Roos, "Corrosion inhibition of C38 steel in 1M hydrochloric acid medium by alkaloids extract from Oxandra asbeckii plant," Corrosion Science, vol. 53, no. 2, pp. 687–695, 2011. ·

32. M. Shyamala and P. K. Kasthuri, "The inhibitory action of the extracts of Adathoda vasica, Eclipta alba, and Centella asiatica on the corrosion of mild steel in hydrochloric acidMedium: a comparative study,"International Journal of Corrosion, vol. 2012, Article ID 852827, 13 pages, 2012. ·

33. M. Shyamala and P. K. Kasthuri, "A comparative study of the inhibitory effect of the extracts of Ocimum sanctum, Aegle marmelos, and Solanum trilobatum on the corrosion of mild steel in hydrochloric acid medium," International Journal of Corrosion, vol. 2011, Article ID 129647, 11 pages, 2011. ·

34. M. Lebrini, F. Robert, and C. Roos, "Inhibition effect of alkaloids extract from Annona squamosa plant on the corrosion of C38 steel in normal hydrochloric acid medium," International Journal of Electrochemical Science, vol. 5, no. 11, pp. 1698–1712, 2010.

35. N. O. Eddy and A. O. Odiongenyi, "Corrosion inhibition and adsorption properties of ethanol extract ofHeinsia crinata on mild steel in H_2SO_4," Pigment and Resin Technology, vol. 39, no. 5, pp. 288–295, 2010. ·

36. E. E. Oguzie, C. K. Enenebeaku, C. O. Akalezi, S. C. Okoro, A. A. Ayuk, and E. N. Ejike, "Adsorption and corrosion-inhibiting effect of Dacryodis edulis extract on low-carbon-steel corrosion in acidic media," Journal of Colloid and Interface Science, vol. 349, no. 1, pp. 283–292, 2010. ··

37. R. Saratha and V. G. Vasudha, "Emblica Officinalis (Indian Gooseberry) leaves extract as corrosion inhibitor for mild steel in 1N HCL medium," E-Journal of Chemistry, vol. 7, no. 3, pp. 677–684, 2010.

38. S. Subhashini, R. Rajalakshmi, A. Prithiba, and A. Mathina, "Corrosion mitigating effect of Cyamopsis Tetragonaloba seed extract on mild steel in acid medium," E-Journal of Chemistry, vol. 7, no. 4, pp. 1133–1137, 2010.

39. M. A. Quraishi, "Investigation of some green compounds as corrosion and scale inhibitors for cooling systems," Corrosion, vol. 55, no. 5, pp. 493–497, 1999.

40. A. Minhaj, P. A. Saini, M. A. Quraishi, and I. H. Farooqi, "A study of natural compounds as corrosion inhibitors for industrial cooling systems," Corrosion Prevention and Control, vol. 46, no. 2, pp. 32–38, 1999.

41. I. H. Farooqi, M. A. Quraishi, and P. A. Saini, "Corrosion prevention of mild steel in 3% NaCl water by some naturally-occurring substances," Corrosion Prevention and Control, vol. 46, no. 4, pp. 93–96, 1999.

42. I. H. Farooqi, M. A. Nasir, and M. A. Quraishi, "Environmentally-friendly inhibitor formulations for industrial cooling systems," Corrosion Prevention and Control, vol. 44, no. 5, pp. 129–134, 1997.

43. M. A. Quraishi, A. Singh, V. K. Singh, D. K. Yadav, and A. K. Singh, "Green approach to corrosion inhibition of mild steel in hydrochloric acid and sulphuric acid solutions by the extract of Murraya koenigii leaves," Materials Chemistry and Physics, vol. 122, no. 1, pp. 114–122, 2010. ·

44. A. Singh, I. Ahamad, V. K. Singh, and M. A. Quraishi, "Inhibition effect of environmentally benign Karanj (Pongamia pinnata) seed extract on corrosion of mild steel in hydrochloric acid solution,"Journal of Solid State Electrochemistry, vol. 15, pp. 1087–1097, 2011. ·

45. A. Singh, V. K. Singh, and M. A. Quraishi, "Aqueous extract of Kalmegh (Andrographis paniculata) leaves as green inhibitor for mild steel in hydrochloric acid solution," International Journal of Corrosion, vol. 2010, Article ID 275983, 10 pages, 2010. ·

46. A. Singh, V. K. Singh, and M. A. Quraishi, "Effect of fruit extracts of some environmentally benign green corrosion inhibitors on corrosion of mild steel in hydrochloric acid solution," Journal of Materials and Environmental Science, vol. 1, no. 3, pp. 162–174, 2010.

47. A. Singh, V. K. Singh, and M. A. Quraishi, "Inhibition effect of environmentally benign Kuchla (Strychnos nuxvomica) seed extract on corrosion of mild steel in hydrochloric acid solution," Rasayan Journal of Chemistry, vol. 3, pp. 811–824, 2010.

48. A. Singh, I. Ahamad, D. K. Yadav, V. K. Singh, and M. A. Quraishi, "The effect of environmentally benign fruit extract of Shahjan (Moringa oleifera) on the corrosion of mild steel in hydrochloric acid solution," Chemical Engineering Communications, vol. 199, no. 1, pp. 63–77, 2012. ·

49. C. Y. Chao, L. F. Lin, and D. D. Macdonald, "A point defect model for anodic passive films," Journal of the Electrochemical Society, vol. 128, no. 6, pp. 1187–1194, 1981.

50. I. M. Ritchie, S. Bailey, and R. Woods, "Metal-solution interface," Advances in Colloid and Interface Science, vol. 80, no. 3, pp. 183–231, 1999. ·

51. G. N. Mu, T. P. Zhao, M. Liu, and T. Gu, "Effect of metallic cations on corrosion inhibition of an anionic surfactant for mild steel," Corrosion, vol. 52, no. 11, pp. 853–856, 1996.

Chapter 7

The Development of a Mathematical Model for The Prediction of Corrosion Rate Behaviour for Mild Steel in 0.5 M Sulphuric Acid

I. Y. Suleiman[1], O. B. Oloche[2], and S. A. Yaro[3]

[1]Department of Metallurgical Engineering, Waziri Umaru Federal Polytechnic, Birnin Kebbi, Nigeria

[2]Department of Mechanical Engineering, University of Abuja, Abuja, Nigeria

[3]Department of Metallurgical and Materials Engineering, Ahmadu Bello University, Zaria, Nigeria

ABSTRACT

The effect of varying temperature, concentration, and time on the corrosion rate of mild steel in 0.5 M H_2SO_4 acid with and without (wild jute tree) grewia venusta plant extract has been investigated by weight loss. The temperature, concentration of inhibitor and time were varied in the range of 0–10% v/v at 2% v/v interval, 30–70°C at 20°C interval, and 45–270 minutes at 45 minutes interval respectively. Scanning electron microscope was used to analyze the morphology of the sample surface. Linear regression equation and analysis of variance (ANOVA) were employed to investigate the influence of process parameters on the corrosion rate of the samples. The predicted corrosion rate of the samples was found to lie close to those experimentally observed ones. The confirmation of the experiment conducted using ANOVA to verify the optimal testing parameters shows that the increase in inhibitor concentration above 2% v/v and time would reduce the corrosion rate. The results also showed that the increase in temperature would also increase the corrosion rate greatly and that the plant extract was very effective for the corrosion inhibition of mild steel in acidic medium.

INTRODUCTION

Steels are the most extensively used structural materials in industry. Mild steel is the most versatile general purpose material due to its good mechanical strength, easy fabricability, formability and weldability, abundance and low cost [1].

In corrosive environments, mild steel structures can be saved by coating and/or cathodic protection. The use of inhibitors is one of the most practical methods for protection against corrosion and prevention of unexpected metal dissolution and acid consumption, especially in acid solutions. Different organic and inorganic compounds have been studied as inhibitors to protect metals from corrosive attack [2].

Such compounds can adsorb onto the metal surface and block the active surface sites, thus reducing the corrosion rate. Although many synthetic compound show good anticorrosive activity, most of them are highly toxic to both human beings and the environment [3], and they are often expensive and non-biodegradable. Thus, the use of natural products as corrosion inhibitors has become a key area of research because plant extracts are viewed as an incredibly rich source of naturally synthesized chemical compounds that are biodegradable in nature and can be extracted by simple procedures with low cost.

Corrosion of mild steel and its alloys in different acid media have been extensively studied [4–6]. Recently considerable interest has been generated in the use of nitrogen, oxygen and sulphur containing organic compounds as corrosion inhibitor for mild steel in different acids [7–9].

In this work, green wild jute tree (Grewa venusta) extract was used as inhibitor. Wild jute tree from the bark ofGrewa venusta, is a shrub or small tree to 10.5 m tall called wild jute in English, Dargaza in Hausa and belong to Tiliaceae family. It is used as fibre and the phytochemical analysis showed that both the Leaves and bark contain many compounds, such as polysaccharides, volatile oils, vitamins, tannins, minerals, alkaloids (e.g., caffeine) and polyphenols (catechins and flavonoids). It is found in the northern part of the country [10].

Thermodynamics and kinetics are useful parameters for analyzing systems undergoing chemical reactivity. Corroding systems are not in equilibrium and therefore thermodynamic calculation cannot be applied. Hence, from the engineering point of view, the major interest is the kinetics or rate of corrosion [11].

In this work, wild jute tree (Grewa venusta) extract was used as inhibitor. Wild jute tree from the bark ofGrewa venusta is a shrub or small tree to 10.5 m tall called wild jute and belongs to Tiliaceae family. It is used as fibre and the phytochemical analysis showed that both the leaves and bark contain many compounds, such as polysaccharides, volatile oils, vitamins, tannins, minerals, alkaloids (e.g., caffeine), and polyphenols (catechins and flavonoids). It is found in the northern part of the country [10].

EXPERIMENTAL PROCEDURE

Preparation of Mild Steel Specimen

Mild steel rods were mechanically cut into cylindrical shape of 20 mm by 10 mm with the following chemical compositions: 0.16% C, 0.38% Mn, 0.18% Si, 0.035% S, 0.034% P and the remainder Fe. The specimen were polished mechanically with emery papers of 80 to 800 grades and subsequently degreased and stored in the desiccators to avoid re-oxidation. Weight of the samples was taken before and after the test.

Preparation of the Plant Extract

The leaves of the plant Grewa venusta was taken and cut into small pieces and they were dried in an air for three days and ground well into powder. The refluxed solution using ethanol was then filtered and the concentration of the stock solution is expressed in terms of (% v/v). From the stock solution, 2–10% v/v concentration of the extract was prepared using 0.5 M sulphuric acid. Similar kind of preparation has been reported in studies using aqueous plant extracts in the recent years [12, 13].

Weight Loss Method

The pretreated specimens' initial weights were noted and were immersed in the experimental solution (in triplicate) with the help of glass hooks at 30°C for a period of 270 minutes. The experimental solution used was 0.5 M H_2SO_4 in the absence and presence of various concentrations of the plant extract. After 270 minutes, the specimens were taken out, washed thoroughly with distilled water, dried completely and their final weights were noted. From the initial and final weights of the specimen, the loss in weight was calculated and tabulated. From the weight loss, the corrosion rate (mpy), inhibition efficiency (%) and surface coverage ( ) of plant extract were calculated using the formula,

$$CR = \frac{534W}{DAT} \ (\mathrm{mpy}),$$

(1)

where W—Weight loss in milligrams (mg), D—Density in grams per cubic centimeter (g/cm³), A—Area of the specimen exposed in square inches (in²) and T—Time of immersion in hours (h)

$$(IE\%) = \left[1 - \frac{CR_i}{CR_l}\right] \times 100,$$

(2)

$$\text{surface coverage } (\theta) = \left[1 - \frac{CRi}{CRl} \right], \tag{3}$$

where CRi and CRl are corrosion rates in the absence and presence of the inhibitors.

Characterization of the Coupons

A Philips model XL30SFEG scanning electron microscope with an energy dispersive X-ray analyzer attached was used in this study. It is a highly-resolution field emission scanning electron microscope with analytical capability. The surface analyses of the coupons before and after corrosion were analyzed for the morphology and the inhomogeneity in the chemical composition. The scanning electron microscope (SEM) was equipped with Energy Dispersive X-ray Spectrometry (EDS) [14].

Development of Mathematical Model

The experimentations were conducted as per standard L8 orthogonal array, so as to investigate which corrosion control parameters significantly affects the corrosion rate and the independently controllable predominant process parameters considered for the investigation are temperature, concentration and time. Two levels of each of the three factors were used for the statistical analysis. The levels for the three factors are entered in Table 1 and the treatment combinations for the two levels and three factors are tabulated in Table 2.

Table 1: Percentage of crude chemical constituents in the plants investigated

S/No.	Plants	Alkaloids	Tannins	Saponins	Flavonoids	Steroids	Volatile
1	A.S	1.43 ± 0.32	15.25 ± 0.11	3.67 ± 0.33	0.00	0.65 ± 0.12	0.65 ± 0.24
2	A.T	1.12 ± 0.39	13.45 ± 0.34	3.33 ± 0.21	0.98 ± 0.33	0.45 ± 0.11	0.86 ± 0.56
3	G.V	1.21 ± 0.45	14.65 ± 0.52	2.99 ± 0.87	0.96 ± 0.59	0.43 ± 10	0.77 ± 0.21

Table 2: Result of corrosion rate without and with inhibitor at different temperature and time

Inhibitor concentration (% v/v)	Mild steel corrosion rate (mpy)			Time (mins)
	30°C	50°C	70°C	
0	51.71	62.41	70.43	135
2	23.26	28.72	35.20	
4	17.06	21.84	27.16	
6	14.47	19.35	25.35	
8	11.94	17.47	21.14	
10	11.02	14.09	18.61	
0	49.81	60.23	69.11	180
2	16.28	23.08	28.81	
4	9.93	12.69	17.23	
6	8.61	11.88	14.71	
8	6.92	10.82	13.15	
10	7.18	9.98	12.74	
0	43.63	56.41	63.60	270
2	10.57	14.54	17.95	
4	8.29	10.72	13.35	
6	6.83	10.35	12.50	
8	6.61	8.95	10.27	
10	6.73	8.79	10.23	

The model equation was obtained by representing the corrosion rate value by CR, the response function can be expressed by equation below:

$$CR = f(A, B, C),$$

(4)

where A is the temperature, B is the inhibitor and C is the time. The model selected includes the effects of main variables first order and second-order interactions of all variables. Hence the general model is written as [11, 15]

$$CR = \beta_0 + \beta_1 A + \beta_2 B + \beta_3 C + \beta_4 AB + \beta_5 AC + \beta_6 BC + \beta_7 ABC, \qquad (5)$$

where $_0$ is average response of CR and β_1, β_2, β_3, β_4, β_5, β_6, and β_7 are coefficients associated with each variable A, B, and C and interactions. The test results were recorded against the standard order of sequence as shown in Table 3. The sum of squares for main and interaction effects was calculated using Yates algorithm. The significant factors (main and interaction) were identified by analysis of variance (ANOVA) technique [16].

Table 3: Result of apparent activation energy for the absence and presence of inhibitor at different exposure time

Inhibitor concentration (% v/v)	Mild steel (apparent activation energy, KJ/mol)	
		Time (mins)
0	2.55	135
2	4.77	
4	5.73	
6	10.19	
8	10.21	
10	10.95	
0	2.45	180
2	4.62	
4	5.65	
6	10.02	
8	10.15	
10	10.75	
0	2.35	270
2	4.45	
4	5.55	
6	9.96	
8	10.02	
10	10.44	

RESULTS AND DISCUSSION

Results

In order to compare the factors that influence corrosion rate of mild steel in 0.5 M sulphuric acid, the corrosion rate of the experimental specimens immersed in the corrosive reagent with and without inhibitor at varied temperature and time were determined using (1) above were shown in Figures 1, 2, 3, 4, 5, 6, and 7.

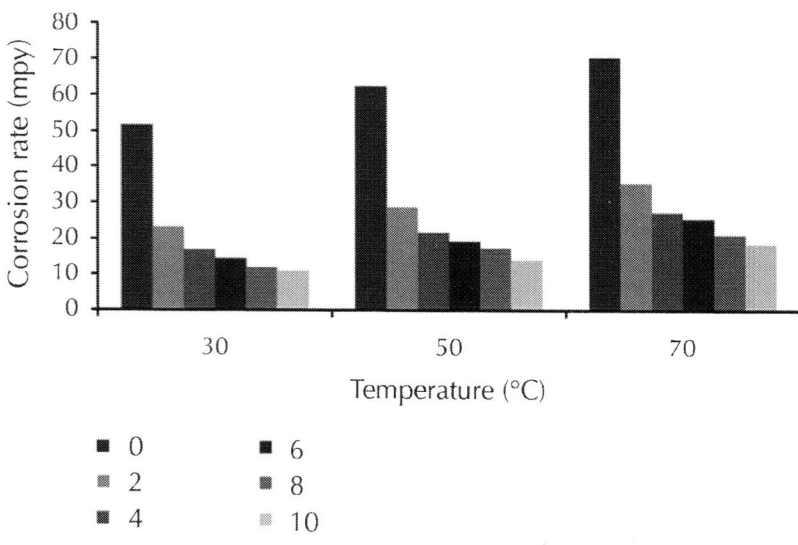

Figure 1: Variation of corrosion rate with temperature after 135 minutes of exposure time.

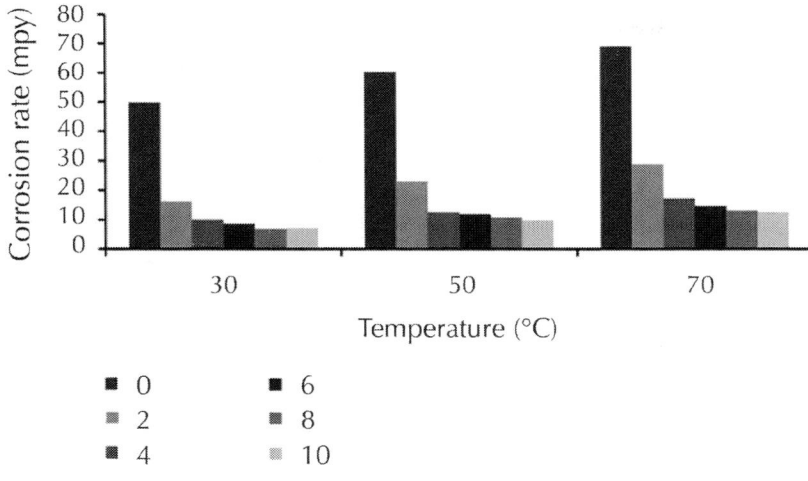

Figure 2: Variation of corrosion rate with temperature after 180 minutes of exposure time.

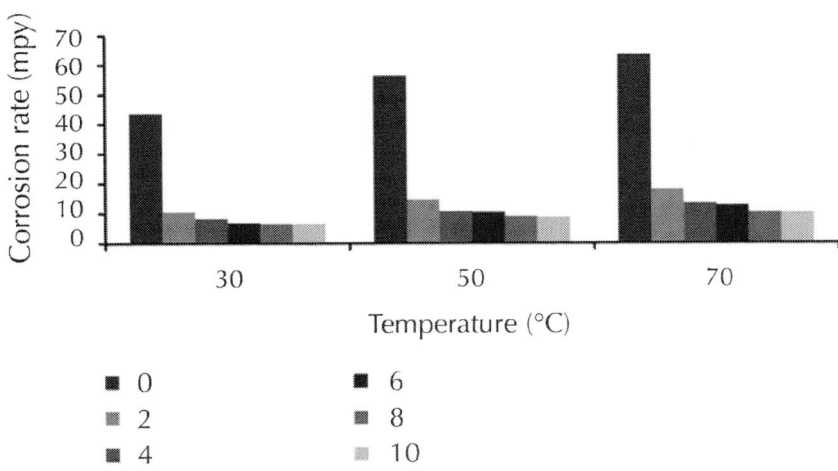

Figure 3: Variation of corrosion rate with temperature after 270 minutes of exposure time.

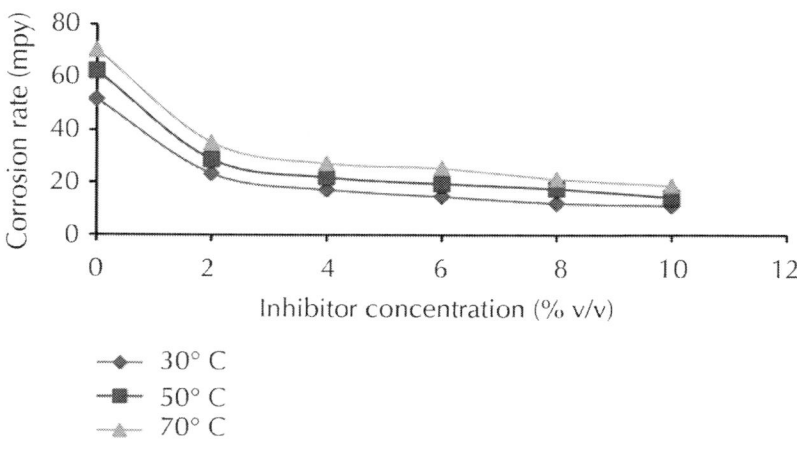

Figure 4: Variation of corrosion rate with inhibitor concentration after 135 minutes exposure time.

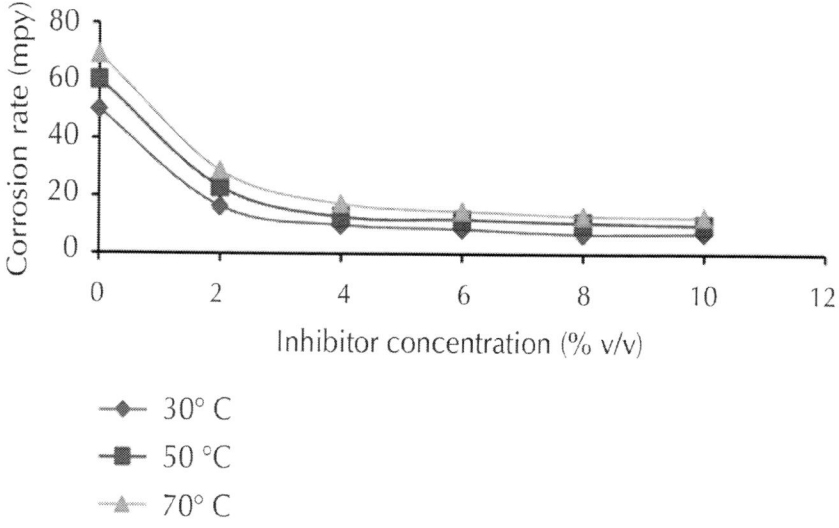

Figure 5: Variation of corrosion rate with inhibitor concentration after 180 minutes exposure time.

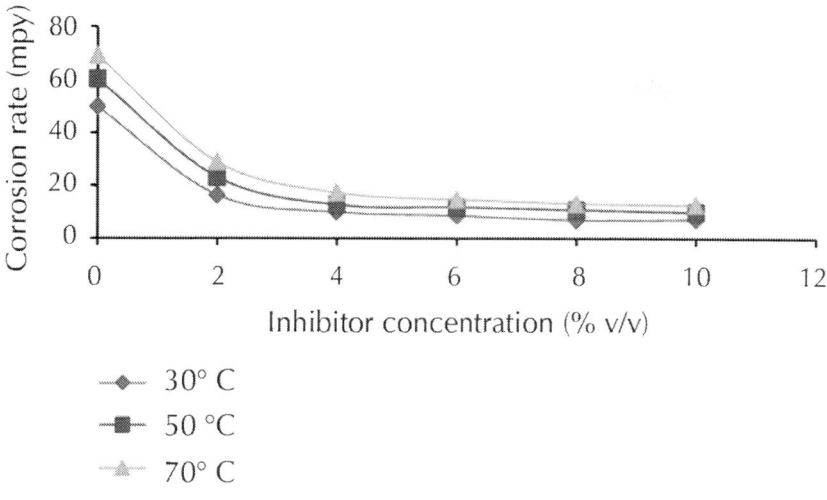

Figure 6: Variation of corrosion rate with inhibitor concentration after 270 minutes exposure time.

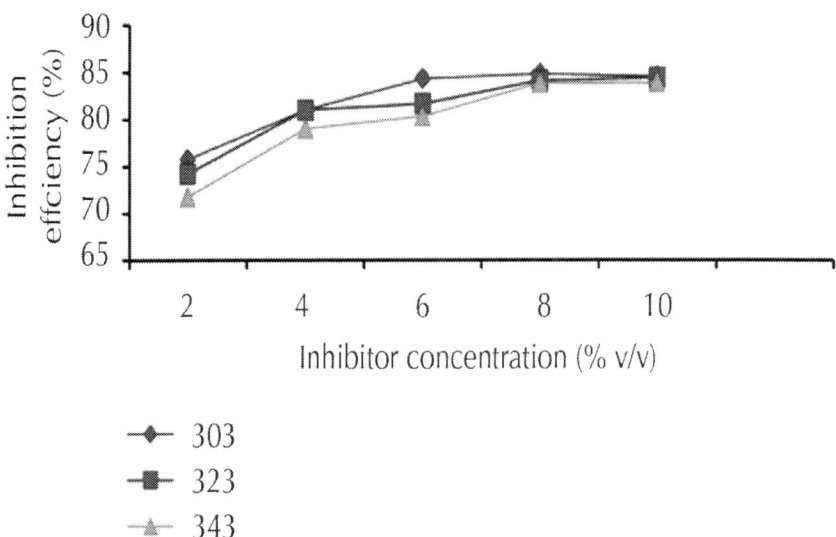

Figure 7: Variations of IE % against concentration of G.V after 270 minutes of exposure at 30, 50, and 70°C.

Kinetics Studies

Although, kinetics models are useful tool to discuss the mechanism of corrosion inhibition of grewa venusta. Arrhenius equation was used to determine the corrosion rate using the expression in (4) and also presented in Figures 8-9

$$A = A_o e^{-Ea/RT}.$$

(6)

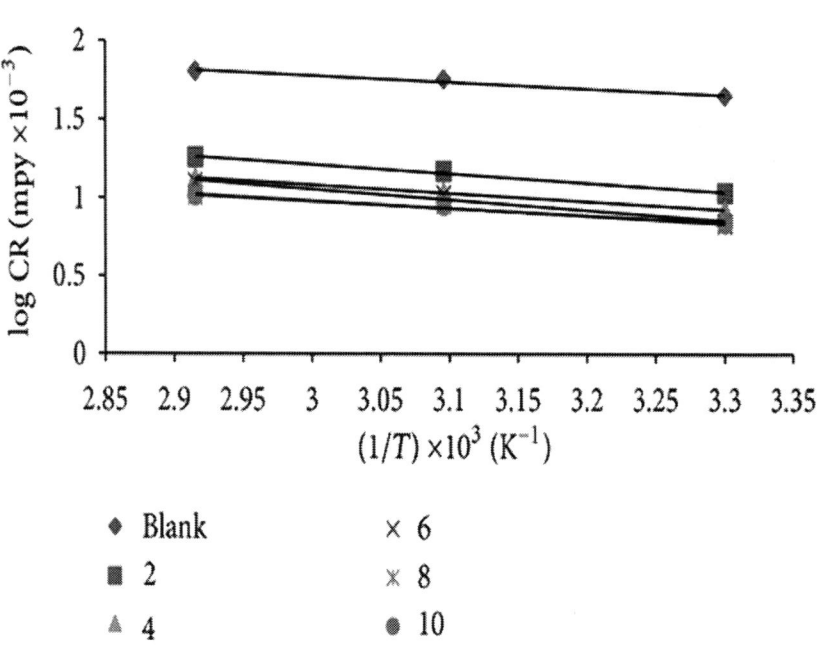

Figure 8: logCR against 1/Temperature for mild steel dissolution process in 0.5 M H_2SO_4 containing different concentrations of Grewa venusta at 30, 50, and 70°C.

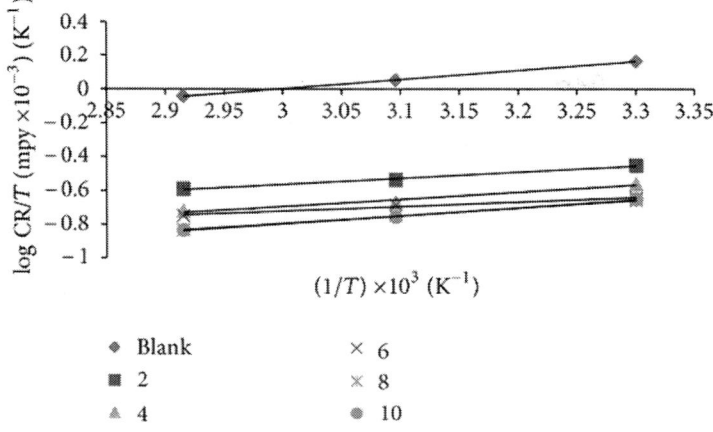

Figure 9: logCR/T against 1/Temperature for mild steel dissolution process in 0.5 M H_2SO_4 containin different concentrations of grewa venusta at 30, 50, and 70°C.

The logarithm of A could be represented as a linear equation given below in (5).

SEM/EDS

The morphology of the polished, with and without inhibitor of grewa venusta were examined and presented in Figures 10–12.

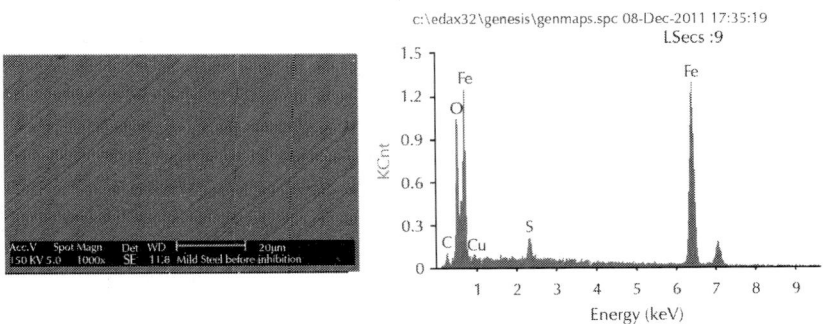

Figure 10: SEM/EDS microstructure of mild steel before corrosion test.

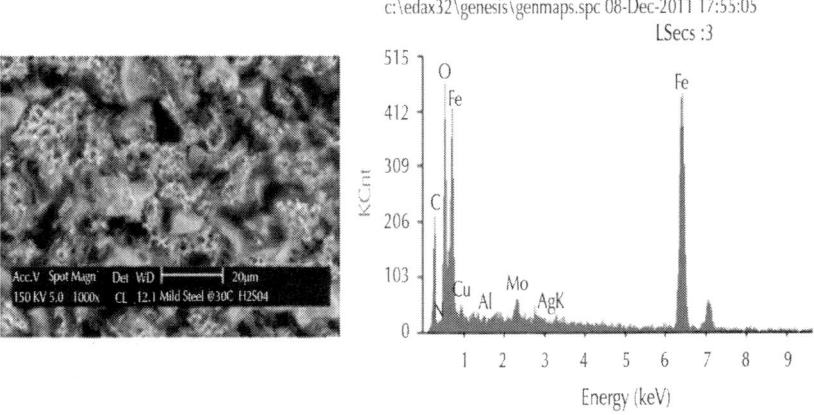

Figure 11: SEM/EDX microstructure of mild steel immersed in 0.5 M H₂SO₄.

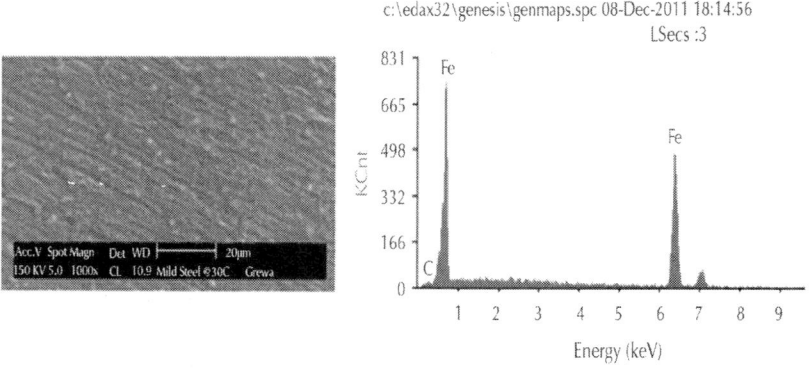

Figure 12: SEM/EDX micrograph of mild steel immersed in 0.5 M H₂SO₄ + 10% v/v G.V.

Development of Model

The results of the statistical model were shown in Tables 7–9 and Figure 13 showed the graph of actual values and predicted values.

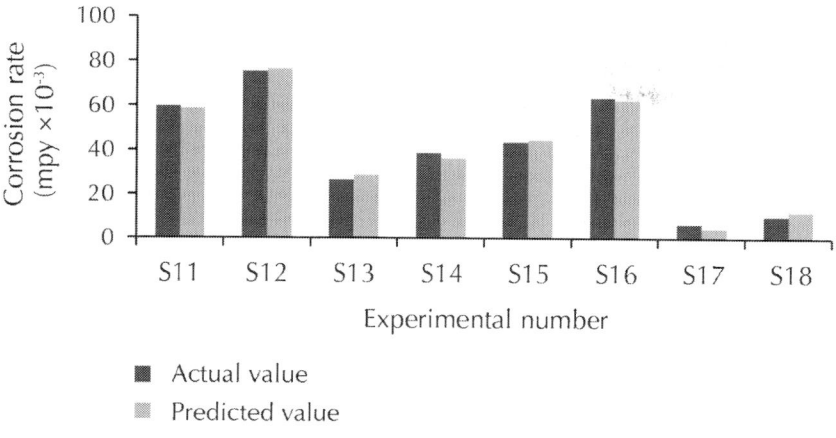

Figure 13: Variation of experimental number with corrosion rate for mild steel with G.V.

DISCUSSION

Figures 1–3 showed the variation of corrosion rate with temperature after 135, 180 and 270 minutes and also in Table 4, it can be seen from these figures and table that the corrosion rate increases with increasing temperature while increase in the plant extract leads to decrease in corrosion rate. From the figures; corrosion rate of 70.43, 69.11 and 63.60 mpy on the specimen immersed in the absence of inhibitor when the temperature was raised from 30 to 70°C at exposure time of 135, 180 and 270 minutes respectively while corrosion rate of 11.02, 7.18 and 6.61 mpy at an experimental temperature of 30°C when the inhibitor concentration were raised from 2 to 10% v/v at an exposure time of 135, 180 and 270 minutes. The figures indicated that temperature, inhibitor and time are significant parameters for the corrosion rate control. Variation of corrosion rate with inhibitor concentration after 135, 180 and 270 minutes of exposure time were shown in Figures 4–6 also confirm that corrosion rates decrease with increase in inhibitor concentration and increase with increase in temperature. This also supported the findings of [8, 9, 17]. Figure 7 shows the variation of inhibition efficiency (IE %) against Concentration of grewa venusta after 270 mins of exposure at 30, 50 and 70°C. From the Table5, it can be concluded that the presence of phytochemical constituents (tannins, alkaloids, saponins, flavonoids) were responsible for the reduction of corrosion rate thereby increasing the efficiency (IE %) of the extract. This is also in support of the findings of [5, 12, 18].

Table 4: Factorial design of the corrosion rate

Factors	Low level	High level
Temperature (A)	30°C	70°C
Inhibitor (B)	0	10% v/v
Time (C)	45 minutes	270 minutes

Table 5: Factorial design of the corrosion process showing treatment combination

Experimental no.	Temperature level	Concentration level	Time level
1	−1	−1	−1
A	+1	−1	−1
B	−1	+1	−1
AB	+1	+1	−1
C	−1	−1	+1
AC	+1	−1	+1
BC	−1	+1	+1
ABC	+1	+1	+1

Coded = −1 (low level), +1 (upper level or high).

Figures 8–9 show the linear regression of logCR and 1/T and logCR/T and 1/T for the specimen immersed in sulphuric acid with and without inhibitor [5, 6]. From the Table 6, the results showed that the apparent activation energy in the absence of inhibitor is lower than that in the presence of the inhibitor which implies physiorption adsorption isotherm [7, 8] and this suggests that the anticorrosion inhibitor is very active for mild steel in acidic medium.

Table 6: Experimental condition and corrosion rate for each shown condition using A.S. extract

Experimental condition	Temperature level (°C)	Concentration level (% v/v)	Time level (mins)	Corrosion rate (mpy × 10−2)
1	30	0	45	43.63
A	70	0	45	75.32
B	30	10	45	38.52
AB	70	10	45	63.60
C	30	0	270	26.48
AC	70	0	270	59.65
BC	30	10	270	6.73
ABC	70	10	270	10.23

Table 7: Analysis of variance (ANOVA) for corrosion rate in the presence of Grewia venusta extract

Source of variation	Sum of squares	Degree of freedom (DF)	Mean square	Fcl=Ms/ Error Ms	F-value	(%)
Main effect						
A	327.42	1	327.42	28.65	0.0332	7.48
B	3209.61	1	3209.61	280.86	0.0035	73.28
C	717.83	1	717.83	62.81	0.0155	16.39
Interaction						
AB	50.50	1	50.50	4.42	0.1703	1.15
BC	51.51	1	51.51	4.51	0.1677	1.18
Residual	22.86	2	11.43			0.52
Cor. Total	4379.73	7				100

Table 8: Effect of the variables at 95% confidence level for Grewia venusta extract

Factor	Coefficient estimate	Degree of freedom	Standard error	95% CI low	95% CI high

Intercept	40.52	1	1.20	35.58	45.66
Temperature (A) 1.00	6.40	1	1.20	1.26	11.54
Inhibitor (B) 1.00	−20.03	1	1.20	−25.17	−14.89
Time (C) 1.00	−9.47	1	1.20	−14.61	−4.33
AB 1.00	−2.51	1	1.20	−7.65	2.63
BC 1.00	−2.54	1	1.20	−7.68	2.60

Table 9: Comparison of the actual and the predicted result for mild steel using Grewia venusta

Exp. no.	Temperature (°C)	Inhibitor (% v/v)	Time (mins)	Corrosion rate (mpy × 10−3)		
				Actual	Predicted	Residual
S11	−1	−1	−1	59.65	58.58	1.07
S12	+1	−1	−1	75.32	76.40	−1.08
S13	−1	+1	−1	26.48	28.62	−2.14
S14	+1	+1	−1	38.52	36.38	2.14
S15	−1	−1	+1	43.63	44.71	−1.08
S16	+1	−1	+1	63.60	62.53	1.07
S17	−1	+1	+1	6.73	4.60	2.13
S18	+1	+1	+1	10.23	12.37	−2.14

Figures 10–12 show different morphologies structures of the coupons of polished, without and with inhibitor. The morphology of the uninhibited surface was altered during corrosion and as expected rough, uneven surface covered and pits and cracks were seen (see Figure 11). However, no pits and cracks were observed in the morphologg of sample with inhibitor (see Figure 12). The protective film formed on the surface of the mild steel was confirmed by SEM studies. Where as in the presence of the optimum extract, mild steel immersed in acidic medium plant extract show the presence of a protective film and

smooth surfaces over the surface of the mild steel in the presence of the inhibitor as shown in Figure 12. This shows that the plant extracts inhibit corrosion of mild steel in acidic solution. This is in line with earlier work of (3, 19).

The results of ANOVA were presented in Table 7, the analysis was evaluated for a confidence level of 95%, that is for significance level of =0.05. It can be observed from the results obtained that inhibitor was the most significant parameter having the highest statistical influence (73.28%) on the corrosion control followed by time 16.39% and temperature 7.48% respectively.

When the -value for the models was less than 0.05, then the parameter or interaction can be considered as statistically significant [19]. From Table 8, it is observed that the temperature (A), inhibitor (B) and time (C) are significant model terms influencing corrosion rate of mild steel. Although the interaction effect of temperature with inhibitor (BC) and inhibitor with time (AB) were considered statistically insignificant since their –F values are greater than 0.05, and hence they are neglected. The coefficient of determination (R^2) is defined as the ratio of the explained variation to the total variation. It is a measure of the degree of fitness. When coefficient of determination R^2 approaches unity, a better response model results and it fits the actual data. The value of R^2 calculated for this model was 0.9165 which means that the developed model has high correlation with the experimental value. It demonstrates that 91.65% of the variability in the data can be explained by this model. Thus, it confirmed that the model provides reasonably good explanation of the relationship between the independent factors and their responses [20]. A multiple linear regression model developed and the effect of 95% confidence levels for the extract was presented in Table 8. A regression equation thus generated establishes correlation between the significant terms obtained from ANOVA, namely, temperature, inhibitor and time. Therefore, it was concluded that the influence of temperature, inhibitor concentration and time on the corrosion rate were statistically significant. The model equation was obtained after calculating each of the coefficients of (7). The developed models equations for the corrosion behaviour of the mild steel in the acidic environment in the presence of the extract can be expressed as:

$$CR\,(G.V) = 40.52 + 6.40A - 20.03B - 9.47C - 2.51AB - 2.54BC. \qquad (8)$$

The results of linear regression model was presented in Table 8 for the extract of G.V which showed that the inhibitor appears to be the most important variable with main effect of −20.03 mpy followed by time (C) with −9.47 mpy and temperature (A) with 6.40 mpy. Similar results have been observed by [11, 21, 22].

The results of linear regression model Table 8 for the extract of G.V showed that the inhibitor appears to be the most important variable with main effect of −20.03 mpy followed by time (C) with −9.47 mpy and temperature (A) with 6.40 mpy. The regression revealed that raising the temperature from 30 to 70°C would result in an increase in the corrosion rates by 6.40 mpy while raising the time from 45 to 270 minutes would result to decrease in corrosion rates by 9.47 mpy and increasing the inhibitor concentrations from 0 to 10% v/v would also result to decrease in the corrosion rates by 20.03 mpy respectively.

The interaction effect of the variables temperature, inhibitor concentration and time are also quite significant and one must take into account these factors for predicting the combined effect of temperature, inhibitor and time on the corrosion rate of the material. The interactive effects are between temperature and inhibitor concentration (BC) and inhibitor concentration and time (AB) are −2.51 and −2.54 mpy. Similar results have been observed by (17, 22,23).

In order to validate the regression model, confirmation test was conducted with parameter levels that were used for analysis. The different parameter levels chosen for the confirmation test are shown in Table 9. Residual variation estimated in (8) for the corrosion rate is in the range of −2.24 to 2.13. The results of the confirmation tests were obtained and comparisons were made between the actual corrosion rate values and the predicted values obtained from the regression models as shown in Table 9. The residual (error) associated with the relationship between the experimental values and the computed values from the regression models for mild steel were very less (less than 4% error). This is in line with findings of [23]. Hence, the regression models developed demonstrated feasible and effective way to predict the corrosion rate of the mild steel. Thus the developed equations can be used to predict corrosion for any combination of factor levels in the specified range. The actual and predicted corrosion rates values are presented in the form of histogram in Figure 13.

CONCLUSIONS

- Experimental data showed that in the presence of different concentration (2–10% v/v), plant extract grewa venusta inhibited the corrosion of mild steel in acidic medium. The inhibition efficiency increased with increase in the extracts concentration and with decrease in temperature leading to a physical adsorption.

- The highest efficiency of 86.47% was observed at the optimum of 8% v/v for grewa venusta extract in the acid solution and the effect of immersion time of the plant extracts at these optimums was attained at 180 minutes immersion time at 30°C which was sufficient for pickling process.

- The value of activation energy Ea calculated from Arrhenius equations revealed that Ea increases in presence of the plant extract in the acid solution and <80 KJ/molK, suggesting that the corrosion inhibition occurred through physical adsorption.

- The SEM micrographs revealed the presence of a protective layers over the metal surface in the presence of the extracts through an adsorption process, hence confirmed the high performance of inhibitive effect of the plant extract.

- EDS results also showed an enhancement of iron in the presence of the extract of grewa venusta.

- ANOVA results revealed that the parameters (A, B, and C) are statistically significant with F-values less than 0.05 and F (%) for the inhibitor grewa venusta are greater than 70 followed by time which is greater than 16 and temperature greater than 7. The interactions exhibited only minor influence and not statistically significant.

- The results obtained by regression equations closely correlate each other which validate the regression equations developed. A good agreement between the predicted and actual corrosion rate was observed.

- The results obtained from the statistical analysis are in good agreement with the experimental findings for the temperature, inhibitor and time. It was found that corrosion rate increases with increasing temperature and decrease with increase in both the inhibitor and time.

- The developed mathematical models can be used to predict the corrosion values in terms of corrosion control process parameters obtained from any combinations within the ranges studied and also employed for optimization of the process parameters of mild steel with respect to corrosion control values.

- The gravimetric method is in a good agreement with the statistical analysis and this improves the validity of the overall results obtained.

REFERENCES

1.   R. Tripathi, A. Chaturvedi, and R. K. Upadhayay, "Corrosion inhibitory effects of some substituted thiourea on mild steel in acid media," Research Journal of Chemical Sciences, vol. 2, no. 2, pp. 18–27, 2012.

2.   S. Rekkab, H. Zarrok, R. Salghi et al., "Green corrosion inhibitor from essential oil of Eucalyptus globulus (Myrtaceae) for C38 steel in sulfuric acid solution," Journal of Materials and Environmental Science, vol. 3, no. 4, pp. 613–627, 2012.

3.   A. Ostovari, S. M. Hoseinieh, M. Peikari, S. R. Shadizadeh, and S. J. Hashemi, "Corrosion inhibition of mild steel in 1 M HCl solution by henna extract: a comparative study of the inhibition by henna and its constituents (Lawsone, Gallic acid, α-d-Glucose and Tannic acid)," Corrosion Science, vol. 51, no. 9, pp. 1935–1949, 2009.

4.   S. A. Abd El-Maksoud, "The effect of organic compounds on the electrochemical behaviour of steel in acidic media. A review," International Journal of Electrochemical Science, vol. 3, no. 5, pp. 528–555, 2008.

5.   A. Begum, S. Harikrishna, I. Khan, and K. Veena, "Enhancement of the inhibitor efficiency of atropine methochloride in corrosion control of mild steel in sulphuric acid," E-Journal of Chemistry, vol. 5, no. 4, pp. 774–781, 2008.

6.   A. A. Rahim and J. Kassim, "Recent development of vegetal tannins in corrosion protection of iron and steel," Recent Patents on Materials Science, vol. 1, no. 3, pp. 223–231, 2008.

7. A. M. Al-Turkustani, S. T. Arab, and R. H. Al-Dahiri, "Aloe plant extract as environmentally friendly inhibitor on the corrosion of aluminum in hydrochloric acid in absence and presence of iodide ions,"Modern Applied Science, vol. 4, no. 5, pp. 105–124, 2010.

8. A. Singh, V. K. Singh, and M. A. Quraishi, "Aqueous extract of kalmegh (Andrographis paniculata) leaves as green inhibitor for mild steel in hydrochloric acid solution," International Journal of Corrosion, vol. 2010, Article ID 275983, 10 pages, 2010.

9. C. A. Loto, R. T. Loto, and A. Popoola, "Synergistic effect of tobacco and kola tree extracts on the corrosion inhibition of mild steel in acid chloride," International Journal of Electrochemcal Science, vol. 6, no. 9, pp. 3830–3843, 2011.

10. B. O. Obadoni and P. O. Ochuko, "Phytochemical studies and comparative efficacy of the crude extracts of some haemostatic plants in edo and delta states of Nigeria," Global Journal of Pure and Applied Sciences, vol. 8, no. 2, pp. 203–208, 2001.

11. O. B. Oloche, S. A. Yaro, and E. G. Okafor, "Analytical correlation between varying corrosion parameters and corrosion rate of Al-4.5Cu/10%ZrSiO$_4$ composite in hydrochloric acid by rare earth chloride," Journal of Alloys and Compounds, vol. 472, no. 1-2, pp. 178–185, 2009.

12. A. M. Abdel-Gaber, B. A. Abd-El-Nabey, and M. Saadawy, "The role of acid anion on the inhibition of the acidic corrosion of steel by lupine extract," Corrosion Science, vol. 51, no. 5, pp. 1038–1042, 2009.

13. G. Ilayaraja, A. R. Sasieekhumar, and P. Dhanakodi, "Inhibition of mild steel corrosion in acidic medium by aqueous extract of tridax procumbens L," E-Journal of Chemistry, vol. 8, no. 2, pp. 685–688, 2011.

14. J. B. Wachtman and R. A. Haber, Ceramic Films and Coatings, Noyes Publications, Park Ridge, NJ, USA, 1993.

15. V. S. Aigbodion, S. B. Hassan, E. T. Dauda, and R. A. Mohammed, "The development of mathematical model for the prediction of ageing behaviour for Al-Cu-Mg/bagasse ash particulate composites," Journal of Minerals and Materials Characterization & Engineering, vol. 9, no. 10, pp. 907–917, 2010.

16. I. Miller and J. E. Freund, Probability and Statistics for Engineers, Prentice Hall India, New Delhi, India, 2001.

17. L. E. Umoru, I. A. Fawehinmi, and A. Y. Fasasi, "Investigation of the inhibitive influence of theobroma cacao and cola acuminata leaves extracts on the corrosion of a mild steel in sea water," Journal of Applied Sciences Research, vol. 2, no. 4, pp. 200–204, 2006.

18. N. O. Obi-Egbedi, K. E. Essien, and I. B. Obot, "Computational simulation and corrosion inhibitive potential of alloxazine for mild steel in 1M HCl," Journal of Computational Methods in Molecular Design, vol. 1, no. 1, pp. 26–43, 2011.

19. S. M. Ross, Introduction to Probability and Statistics for Engineers and Scientists, Elsevier Academic Press, 3rd edition, 2004.

20. S. Venkat Prasat, R. Subramanian, N. Radhika, B. Anandavel, L. Arun, and N. Praveen, "Influence of parameters on the dry sliding wear behaviour of aluminium/fly ash/graphite hybrid metal matrix composites," European Journal of Scientific Research, vol. 53, no. 2, pp. 280–290, 2011.

21. E. E. Oguzie, "Corrosion inhibitive effect and adsorption behaviour of Hibiscus sabdariffa extract on mild steel in acidic media," Portugaliae Electrochimica Acta, vol. 26, no. 3, pp. 303–314, 2008.

22. M. A. Amin, S. S. Abd El-Rehim, E. E. F. El-Sherbini, and R. S. Bayoumi, "Chemical and electrochemical (AC and DC) studies on the corrosion inhibition of low carbon steel in 1.0 M HCl solution by succinic acid-temperature effect, activation energies and thermodynamics of adsorption," International Journal of Electrochemcal Science, vol. 3, no. 2, pp. 199–215, 2008.

23. P. G. Kochure and K. N. Nandurkar, "Mathematical modeling for selection of process parameters in induction hardening of EN8 D steelJournal of Mechanical and Civil Engineering," vol. 1, no. 2, pp. 28–32, 2012.

Copper Corrosion by Atmospheric Pollutants in the Electronics Industry

Benjamin Valdez Salas[1], Michael Schorr Wiener[1], Roumen Zlatev Koytchev+, Gustavo López Badilla[2], Rogelio Ramos Irigoyen[1], Monica Carrillo Beltrán[1], Nicola Radnev Nedev[1], Mario Curiel Alvarez[1], Navor Rosas Gonzalez[2], and Jose María Bastidas Rull[3]

[1]Engineering Institute, Autonomous University of Baja California, Boulevard Benito Juarez y Calle a la Normal S/N, Colonia Insurgentes Este, 21280 Mexicali, BCN, Mexico

[2]Polytechnic University of Baja California, Calle de la Claridad S/N, Colonia Plutarco Elias Calles, 21376 Mexicali, BCN, Mexico

[3]National Center of Metallurgical Research, Avenue Gregorio del Amo 8, 28040 Madrid, Spain

ABSTRACT

Hydrogen sulphide (H_2S) is considered one of the most corrosive atmospheric pollutants. It is a weak, diprotic, reducing acid, readily soluble in water and dispersed into the air by winds when emitted from natural, industrial, and anthropogenic sources. It is a pollutant with a high level of toxicity impairing human health and the environment quality. It attacks copper forming thin films of metallic sulphides or dendrite whiskers, which are cathodic to the metal substrate, enhancing corrosion. H_2S is actively involved in microbially influenced corrosion (MIC) which develops in water, involving sulphur based bacteria, in oxidizing and reducing chemical reactions. H_2S is found in concentrated geothermal brines, in the atmosphere of geothermal fields, and in municipal sewage systems. Other active atmospheric pollutants include SO_x, NO_x, and CO. This investigation reports on the effects of H_2S on copper in microelectronic components of equipment and devices, with the formation of nonconductive films that lead to electrical failures.

INTRODUCTION

The electronics industry is spread out worldwide; it is an important sector in the Mexican economy, representing 80% of industrial companies in the northwest of the country. Their assembly plants are located in three cities: Mexicali, an arid zone, Tijuana, an urban-industrial area, and Ensenada, a marine region on the Pacific Ocean coast, all belonging to the State of Baja California, near the Mexico-USA border. The electronics industry appeared in Mexico during the sixties with the manufacture of electronic products such as radios, audio record and play devices, and televisions. This industry designs and manufactures microelectronic components called microcontrol devices (MCD), integrated with microelectromechanical systems (MEMS). A study was conducted in the indoor areas of three electronics plants in these cities. Copper and its alloys are widely applied in the electric energy, electronics, and semiconductor industries because of their high electrical and thermal conductivity, ductility, and malleability.

Copper is considered a noble metal; it resists attack by oxygen, although some air pollutants, such as H_2S, change its surface properties, even at ambient temperature, forming a thin layer having completely different properties compared with the pure metal surface. This layer lowers catastrophically the adhesion of the soldering alloy or conductive resins and paste, provoking failures of the printed circuit board (PCB) of the microelectronic devices. Compounds such as geerite (Cu_8S_5) are formed on Cu in the presence of H_2S and patinas as

Cu_2O; posnjakite ($Cu_4SO(OH)_6H_2O$), brochantite ($Cu_4SO_4(OH)_6H_2O$), and antlerite ($Cu_3SO_4(OH)_4$) are formed in the presence of humidity [1]. The formation of tarnish films on a copper surface exposed to environments containing atmospheric pollutants and high humidity involves the movement of metallic ions over the surface, away from the metal generating a creep process that increases the contact resistance, leading to electric failures of the electronic devices. Copper sulphidation is a fast process occurring on the metal-gas phase interface impairing the Cu corrosion resistance [2, 3]. This paper presents the corrosion process of Cu exposed to H_2S polluted environments, under varied conditions of temperature and humidity, in the indoor areas of electronics manufacturing plants located in Mexicali, Baja California, Mexico [4].

H_2S, A CORROSIVE, TOXIC POLLUTANT

It is appropriate to report in the context of the present paper about the corrosivity and toxicity of H_2S since this also affects the quality of the environment, human health, and the durability of the engineering materials, which are central issues of modern society. H_2S acts as a pollutant in the indoor areas of manufacturing plants of the electronics industry; it promotes the formation of thin copper sulphide films on PCB surfaces. Recent investigations [5, 6] proved that the main H_2S source is a geothermal field generating underground sources of steam and H_2S located about 40 km south from Mexicali city. To avoid this air pollution and consequent corrosion is impossible without the application of high cost air cleaners [7, 8].

H_2S gas emitted into the atmosphere from additional heavy sources such as municipal sewage causes respiratory diseases and inflammation of the eyes; it has an offensive odor of rotten eggs; therefore, it is easy to detect, even at low concentration of 10 to 30 ppb (parts per billion) in the atmosphere around the geothermal fields.

The corrosion activity of H_2S is evident in

$$Fe + H_2S \longrightarrow FeS + H_2$$

(1)

$$Cu + H_2S \longrightarrow CuS + H_2$$

(2)

EXPERIMENTAL

Air Pollutants Measurements

Data on air pollutants were gathered every five minutes and organized in files for monthly periods. The specialized instruments controlled by the United States Environmental Protection Agency (US-EPA) monitoring air pollution were a chemiluminescence NO_x analyzer, model 42 of Thermo Environmental Instruments Inc., a gas filter CO analyzer model 3000E of Advanced Pollution Instruments Inc., (API), a SO_2 photometric analyzer from Thermo Electron Corporation, and an O_3 analyzer model 400 of API. This electronic instrument has filters to separate dust particles from gases.

The practices recommended in ISO standards for atmospheric corrosion were taken into account. The Cerro Prieto geothermal wells, in the vicinity of Mexicali, emit H_2S into the atmosphere surrounding the fields and the power plants. Other H_2S emissions came from the plant chimney stacks, vapour ducts, noise silencers, and cooling towers, totaling 22,740 t per year [9, 10]. Since the production capacity of Cerro Prieto has not changed over the last 5 years or so, it may be estimated that the same H_2S concentration in the atmosphere, around the geothermal wells, ranges from 10 to 30 ppb. Typical ranges of natural and anthropogenic H_2S under outdoor and indoor conditions are 0.724 ppb and 0.1 to 0.7 ppb, respectively [11, 12].

Corrosion Rate Measurement

Rectangular metallic specimens of Cu with an exposition surface of 6.45 cm^2 were prepared. The specimens were polished to 400 SiC paper, washed, degreased with acetone, dried with hot air, and weighed before being installed in a metallic chamber of exposure under indoor conditions for 1-, 3-, 6-, 12- and 24-month periods. The corrosion rates were determined by applying the gravimetric method according to ASTM G 31 standard method. In order to simulate controlled indoor conditions, the chamber was fabricated with precoated aluminum with a total volume of 0.1 m^3 and conditioned with two air inlet blinds coupled to metallic filters in order to permit the penetration of gases with the flow of air, to prevent the penetration of dust, and to avoid mistakes in the weight loss calculations [13, 14]. The chamber was provided with metallic internal supports to hold the specimens; it was installed on the roof of an electronics manufacturing plant at 10 m above ground level [15, 16].

After each period of exposure, the metallic samples were removed and weighed to obtain the mass gain. The corrosion products morphology was observed with a stereoscope before being removed, cleaned, and reweighed to obtain the mass loss on an analytical balance to the nearest .00001 g of accuracy. Corrosion of Cu surfaces under constant concentration of H_2S and controlled relative humidity (RH) conditions occurred using a closed system consisting of an acrylic sealed chamber with an inlet valve in order to provide a 0.1 ppm H_2S concentration and a ventilator coupled to a humidity generator to provide 80% RH [17, 18].

Surface Examination

The copper specimens exposed to a controlled H_2S environment during short times were analyzed to determine their surface characteristics by a Scanning Electron Microscopy (SEM) applying a JEOL (JEOL, Ltd., Peabody, MA) JSM-6360, coupled with an Energy Dispersive X-ray (EDX) analyzer (AMETEK, Inc., Mahwah, NJ), used for chemical composition analysis.

RESULTS

Corrosion of Copper

The corrosion rate (CR) of Cu specimens exposed in the test chamber during 24 months (Figure 1) demonstrates that the extent of corrosion augments with the exposure time reaching values of $270 \, mg \cdot m^{-2}$ after the two years of exposure at RH values ranged from 15% to 75% and temperatures from 4°C to 45°C, depending on the year season; SO_2 and NO_X were the species with major concentration levels.

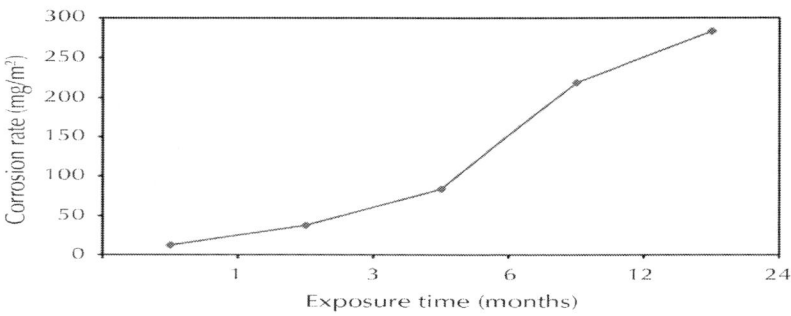

Figure 1: Corrosion rate of copper exposed in the corrosion chamber during 24 months.

Corrosion of Cu in the presence of air, H_2S, and humidity produces oxides and/or sulphides as wet films leading to electrical failures in electronic equipment [10]. On the other hand, corrosion of Cu occurs when the RH overpasses the 80% and the SO_2 concentration is larger than 0.1 ppm. The corrosion behavior also depends on the state of the metal surface: smooth, corrugated, polished, and nonuniform. Sometimes, a Cu_2O film that slows the rate of corrosion gradually dissolves in the presence of an acidic electrolyte constituted by SO_2, and then the corrosion products formed are different due to the surface conditions. At levels of SO_2 greater than the air quality standards, in combination with NO_2 and O_3 at different concentrations in indoor plants, cuprite (Cu_2O) and copper sulphides form, as depicted in Figures 2 and 3.

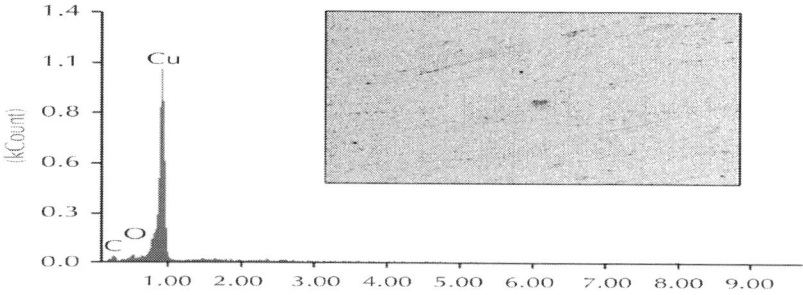

Figure 2: SEM picture for Cu surface after 2 days in 0.3 ppm H_2S/air, magnification 400x.

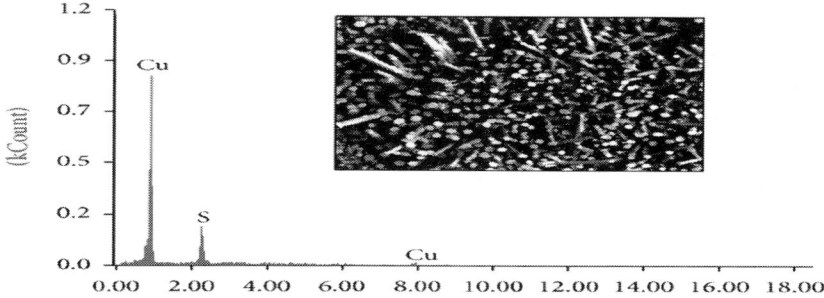

Figure 3: SEM picture for Cu surface after 2 days in 0.3 ppm H_2S/air, magnification 1200x.

Some spectacular structures of hexagonal crystals of Cu sulphide are shown in Figure 3. The SEM results of the exposed samples showed the formation of spots of CuS during the first two days, which grew continuously until all spots united together as large films covering the entire sample surface. The concentration of H_2S in the electronics plant atmosphere was 0.9 ppm; at the Cerro Prieto geothermal field atmosphere it reaches 1.5 ppm and higher concentrations, due to the continuous emission of gases accompanying the produced steam [11].

The EDX analysis performed in different points of the corroded copper surfaces reveals the formation of several sulphides of stoichiometric composition with a general formula Cu_xS, where the values of X vary from 1.6 to 2. The estimated composition of the copper sulphide corrosion products and the corresponding mineral are displayed in Table 1.

Table 1: Copper sulphides in films formed by reaction with H_2S at indoor areas of electronics plants in Mexicali City

Mineral	Chemical composition
Chalcocite	Cu_2S
Djurlcite	$Cu_{1.95}S$
Digenite	$Cu_{1.82}S$
Geerite	$Cu_{1.62}S$

Film Formation Mechanisms

Under conditions of humidity and in contact with air contaminated with H_2S, Cu generates two types of corrosion products: oxides and sulphides according to the following electrochemical reactions.

Oxidation

$$2Cu \longrightarrow 2Cu^{2+} + 4e^- \tag{3}$$

Oxidation, anodic reaction

$$O_2 + 2H_2O + 4e^- \longrightarrow 4OH^- \tag{4}$$

Reduction, cathodic reaction

$$2Cu + O_2 + 2H_2O \longrightarrow 2Cu(OH)_2 \tag{5}$$

Total corrosion reaction

$$2Cu(OH) \longrightarrow Cu_2O + H_2O \tag{6}$$

Hydroxide converts to oxide

Sulphidation

$$Cu \longrightarrow Cu^{2+} + 2e^-$$

(7)

Oxidation, anodic reaction

$$H_2S \longrightarrow H^+ + SH^-$$

(8)

Reduction, cathodic reaction

$$2SH^- \longrightarrow 2S^- + H_2$$

(9)

Reduction, cathodic reduction

$$Cu + H_2S \longrightarrow CuS + H_2$$

(10)

Total sulphidation reaction.

Influence of Atmospheric Pollutants

Mexicali city has highly contaminated air because of the presence of fine dust coming from the desert around, but the gaseous pollutants such as SO_x, NO_x, CO, and O_3 are generated from the diverse industrial activities. Air pollutants penetrate to indoor locations of electronics plants and corrode Cu-made devices. The relative humidity (RH) and temperature reach up to 50% and 30°C during the major part of the year. Table 2 presents the relation between the concentrations of these pollutants and the Mexicali climatic factors. The two most aggressive pollutants are sulphur-containing H_2S and SO_2, both acidic, but one reducing (H_2S) and the other oxidant (SO_2) agents. Their behaviour and corrosivity during the seasons of the year are presented in Table 3.

Table 2: Relation of concentration of air pollutants and climate factors in Mexicali

Climatic factors												
Seasons	Sulfur dioxide (SO$_2$)			Carbon monoxide (CO)			Nitrogen oxides (NO$_x$)			Ozone (O$_3$)		
	RH^a	T^b	C^c	RH^a	T^b	C^c	RH^a	T^b	C^c	RH^a	T^b	C^c
Spring												
Max	49.3	34.8	0.16	39.1	29.3	69	34.3	28.6	0.48	48.3	27.3	0.21
Min	23.3	20.4	0.08	28.6	22.1	4	29.8	19.9	0.01	28.5	15.2	0.06
Summer												
Max	83.8	45.9	0.11	73.9	45.9	57	70.8	39.9	0.54	73.2	44.5	0.37
Min	46.1	23.6	0.03	47.5	29.2	16	44.7	22.4	0.18	44.3	26.5	0.01
Winter												
Max	77.8	27.4	0.50	72.8	24.7	84	69.1	24.6	0.75	87.9	27.7	0.51
Min	16.6	17.8	0.17	39.3	23.9	6	63.2	13.8	0.17	47.2	28.8	0.05

^{a}RH: relative humidity, %, ^{b}T: temperature, °C, and ^{c}C: air pollution concentration (C), ppm.

Table 3: Correlation of corrosion rate, the year season, and air pollutants in indoor conditions of industrial plants

Seasons	Climatic factors							
	Hydrogen sulfide (H$_2$S)				Sulfur dioxide (SO$_2$)			
	RH[a]	T[b]	C[c]	CR[d]	RH[a]	T[b]	C[c]	CR[d]
Spring								
Max	88.8	33.4	0.15	255	85.6	23.2	0.34	176
Min	34.5	17.6	0.09	130	46.7	15.1	0.23	112
Summer								
Max	89.9	42.1	0.14	265	88.2	39.9	0.45	245
Min	38.5	24.3	0.11	181	42.3	28.2	0.18	114
Winter								
Max	87.5	25.6	0.42	382	88.8	22.3	0.67	338
Min	43.2	17.8	0.26	245	38.9	12.3	0.25	136

[a]RH: relative humidity, %, [b]T: temperature, °C, [c]C: air pollution concentration (C), ppm, and [d]CR: corrosion rate, mg/m²·year.

The corrosion data collected, arranged in Tables 2 and 3, correlate values of RH, temperature, and CR provoked by the different atmospheric pollutants. These data were evaluated and displayed using the MATLAB software, a mathematical computing software (MathWorks Inc., USA), to determine the relationship between the environmental factors and the corrosion rate of metals used in the electronics industry. Figure 4 is a particular 3D graph depicting the correlation of the CR of Cu with the RH and the temperature, indicating in circles the maximum and the minimum CR.

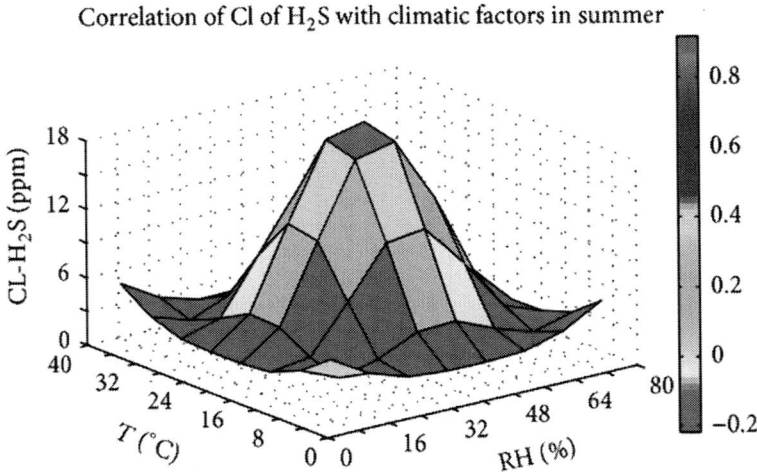

Figure 4: Correlation of copper corrosion with H$_2$S, relative humidity %, and temperature in the electronics industry (summer, 2010).

The maximum CR appears at 88% RH and 16°C and the minimum CR is recorded at 18% RH and 2.0°C, expressing the critical influence of humidity and temperature levels. These levels are controlled inside the electronics plant but sometimes corrosion occurs.

An additional MATLAB graph depicts the influence of frequent industrial pollutants: NO$_x$, SO$_x$, O$_3$, and CO$_x$ on the corrosivity indexes of Cu published by ISO, the International Organization for Standardization [7] (Figure 5).

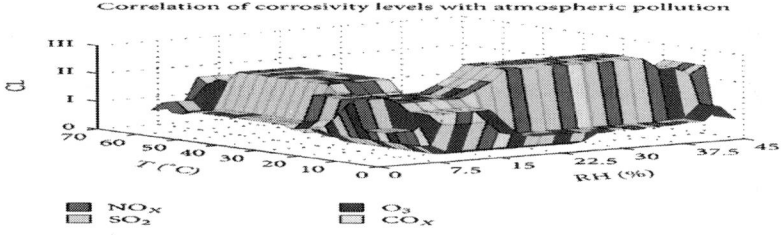

Figure 5: Correlation of climatic factors and pollutants with corrosivity indexes.

Other MATLAB graphs correlate the CR of Cu with the climatic factors and pollutants in summer 2010 in Mexicali (Figure 6) and those in winter 2010 too (Figure 7).

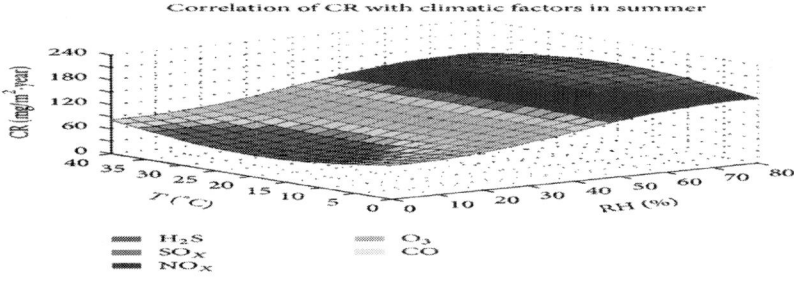

Figure 6: Correlation of CR of copper with climatic factors in summer in Mexicali (2010).

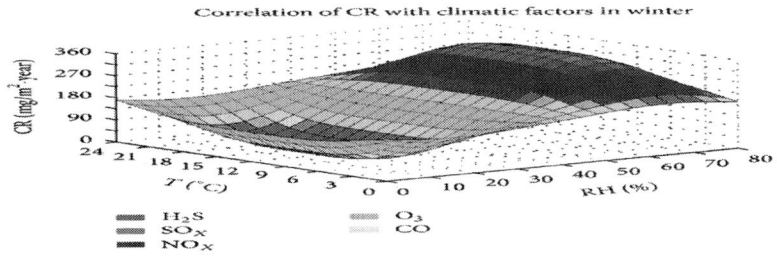

Figure 7: Correlation of CR of copper with climatic factors in winter in Mexicali (2010).

DISCUSSION

The climatic variables and the atmospheric pollutants are the principal factors that enhance the corrosion in indoor conditions of the metals utilized in the electronics industry of the State of Baja California, Mexico. The evaluation of these parameters and their effect on the metals' surfaces demonstrates the relationship of atmospheric corrosion with the damage caused to the electrical connections, resistors, diodes, connectors, and wires of the electrical-electronic equipment. This corrosion damage generates low yielding by electrical failures in the industrial devices and equipment. The maximum and minimum RH, temperature, and CR and the relationship with the air pollutants were analyzed during different seasons of the year. These data were expressed graphically applying the MATLAB software. A study on the correlation of climate factors with the function of electronic test equipment installed inside a clean room of an electronics plant in the city of Tijuana was conducted; it was done at various levels of RH and temperature, relating them to the electric current circulating in the test equipment to indicate the correct or incorrect state of the microcomponents. They include electronic components such as transistors, capacitors, coils, resistors, and diodes that are assembled in the semiconductor wafer based on a silicon microelectronic board. The bad air quality in Tijuana should be attributed to the last cold winters that have caused more burning of fossil fuels to supply electricity for urban-heating systems, increasing the emissions of corrosive pollutants, worsening the smog which comprises fire smoke, car exhaust gases, and dust, all trapped in the air mist. The health impact is vast too, raising the number of people suffering from respiratory illness that triggers heart and asthma attacks. The city of Ensenada is located on the coast of the Pacific Ocean, in the northwest of Mexico in a marine region. It has a tropical marine climate with cold winter mornings around 5°C and 35°C in the summer; RH is around 20% to 80%, varying during the seasons of the year. The climate factors analyzed were humidity, temperature, winds, and rains to determine the time of wetness (TOW), a critical factor in the determination of CR and its extent.

Santa Ana winds (SAW) constitute a climatic phenomenon that alters the atmospheric conditions; they originate in the Santa Ana canyon, in the Mojave desert, which cause fast changes in the climate conditions in southwest California and northwest Baja California. SAW form when the desert becomes cooler, usually during the autumn and spring seasons; rising temperatures, humidity, and meteorological conditions influence the indoor environment of the electronics industry. The principal air corrodent encountered in Ensenada is the NaCl aerosols from the sea, in addition to CO, NO_x and SO_2 from traffic vehicles, power stations, and industrial and landfill emissions which increase atmospheric corrosivity [13, 14].

CONCLUSIONS AND RECOMMENDATIONS

A study was conducted during a span of two years on the corrosion of metallic materials used in the manufacture of electronic devices and equipment. The gaseous air pollutants, for example, H_2S, SO_x, NO_x, and CO, generated by the geothermal fields, the electricity industry, and the motor vehicles burning fossil fuels, lead to the appearance of corrosion on the metals' surfaces. Copper suffers in particular due to the attack by sulphur-containing pollutants: H_2S and SO_x forming copper sulphides and oxides that impair their electrical conductivity properties. The installation and effective maintenance systems, to clean and control the contaminated air that infiltrates into the electronics production rooms such as filters would prevent and/or minimize this atmospheric corrosion. Encapsulation and hermetic sealing of microcomponents prevent a reaction with the pollutants. Removal of moisture by efficient air conditioning and continuous maintenance of an optimum environmental condition of the indoor areas of electronics plants avoids and/or mitigates corrosion.

REFERENCES

1. B. Valdez, M. Schorr, M. Quintero et al., "Corrosion and scaling at Cerro Prieto geothermal field," Anti-Corrosion Methods and Materials, vol. 56, no. 1, pp. 28–34, 2009.

2. L. Veleva, B. Valdez, G. Lopez, L. Vargas, and J. Flores, "Atmospheric corrosion of electro-electronics metals in urban

desert simulated indoor environment," Corrosion Engineering Science and Technology, vol. 43, no. 2, pp. 149–155, 2008.

3.   J. F. Flores and S. B. Valdez, "Cabina de simulación de corrosion para la industria electrónica en interior," Ingenieros, vol. 6, no. 21, 2003 (Spanish).

4.   ASTM, "Standard practice for conducting atmospheric corrosion test on metals," ASTM G50-76, ASTM, West Conshohocken, Pa, USA, 2003.

5.   G. Lopez, B. Valdez, and M. Schorr, "Spectroscopy analysis of corrosion in the electronics industry influenced by Santa Ana Winds in marine environments of Mexico," in Indoor and Outdoor Air Pollution, INTECH, 2011.

6.   "Corrosion of metals and alloys. Classification of low corrosivity of indoor atmospheres: determination and estimation attack in indoor atmospheres," ISO 11844-1, ISO, Geneva, Switzerland, 2005.

7.   "Corrosion of metals and alloys. Classification of low corrosivity of indoor atmospheres: determination and estimation of indoor corrosivity," ISO 11844-2, ISO, Geneva, Switzerland, 2006.

8.   A. Moncmanova, Environmental Deterioration of Materials, WIT Press, 2007.

9.   L. B. Gustavo, Caracterización de la corrosión en materiales metálicos de la industria electrónica en Mexicali, B.C. [Tesis de doctorado], UABC, Instituto de Ingeniería, Mexicali, México, 2008.

10.   G. López, H. Tiznado, G. Soto, W. de la Cruz, B. Valdez, and R. M. Schorr Zlatev, "Corrosión de dispositivos electrónicos por contaminación atmosférica en interiores de plantas de ambientes áridos y marinos," Revista Nova Scientia, vol. 3, no. 1, 2010.

11.   G. López, H. Tiznado, G. S. Herrera et al., "Use of AES in corrosion of copper connectors of electronic devices and equipments in arid and marine environments," Anti-Corrosion Methods and Materials, vol. 58, no. 6, pp. 331–336, 2011.

12.   M. Reid, J. Punch, C. Ryan et al., "Microstructural development of copper sulfide on copper exposed to humid H_2S," Journal of the Electrochemical Society, vol. 154, no. 4, pp. C209–C214, 2007.

13. B. Valdez, M. Schorr, R. Zlatev et al., "Corrosion control in industry," in Environment and Industrial Corrosion, Practical and Theoretical Aspects, INTECH, 2012.

14. S. B. Valdez, W. M. Schorr, B. G. Lopez et al., "H_2S pollution and its effect on corrosion of electronic components," in Air Quality-New Perspective, INTECH, 2012.

15. B. G. Lopez, S. B. Valdez, W. M. Schorr, and G. C. Navarro, "Microscopy and spectroscopy of MEMS used in the electronic industry of Baja California region Mexico," in Air Quality-New Perspective, INTECH, 2012.

16. B. G. Lopez, S. B. Valdez, K. R. Zlatev, P. J. Flores, B. M. Carrillo, and W. M. Schorr, "Corrosion of metals at indoor conditions in the electronics manufacturing industry," Anti-Corrosion Methods and Materials, vol. 54, no. 6, pp. 354–359, 2007.

17. J. Smith, Z. Qin, F. King, L. Werme, and D. W. Shoesmith, "Sulfide film formation on copper under electrochemical and natural corrosion conditions," Corrosion, vol. 63, no. 2, pp. 135–144, 2007.

18. K. Demirkan, G. E. Derkits Jr., D. A. Fleming et al., "Corrosion of Cu under highly corrosive environments," Journal of the Electrochemical Society, vol. 157, no. 1, pp. C30–C35, 2010.

Chapter 9

Corrosion Protection of Steels by Conducting Polymer Coating

Toshiaki Ohtsuka

Faculty of Engineering, Hokkaido University, Kita 13-jo, Nishi 8-chome, Kita-ku, Sapporo 060-8628, Japan

ABSTRACT

The corrosion protection of steels by conducting polymer coating is reviewed. The conducting polymer such as polyaniline, polypyrrole, and polythiophen works as a strong oxidant to the steel, inducing the potential shift to the noble direction. The strongly oxidative conducting polymer facilitates the steel to be passivated. A bilayered PPy film was designed for the effective corrosion protection. It consisted of the inner layer in which phosphomolybdate ion, $PMo_{12}O_{40}^{3-}$ (PMo), was doped and the outer layer in which dodecylsulfate ion (DoS) was doped. The inner layer stabilized the passive oxide and the outer possessed anionic perm-selectivity to inhibit the aggressive anions such as chloride from penetrating through the PPy film to the substrate steel. By the bilayered PPy film, the steel was kept passive for about 200 h in 3.5% sodium chloride solution without formation of corrosion products.

INTRODUCTION

Since the investigation of Shirakawa et al. on conducting polyacetylene, various applications of conducting polymer have been reported [1]. Utilization of the conducting polymer for corrosion protection coating is one of these applications, and many papers have been presented in the last decade. Preparation of polyacetylen was made by oxidation in gaseous phase; however, at present, the conducting polymers such as polyaniline (PAni), polypyrrole (PPy), and polythiophen (Pthio) in Figure 1 for the corrosion protection have been prepared by electrochemical oxidation in liquid phase.

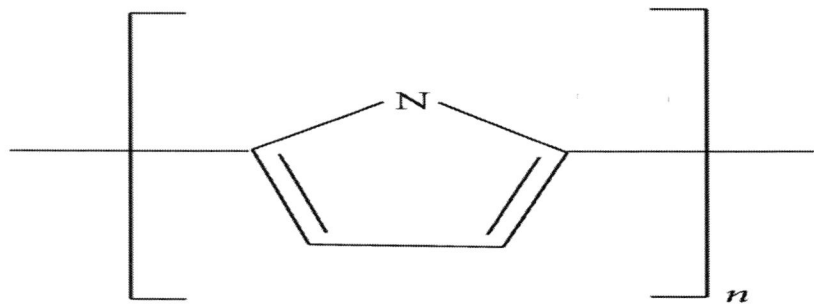

(a)

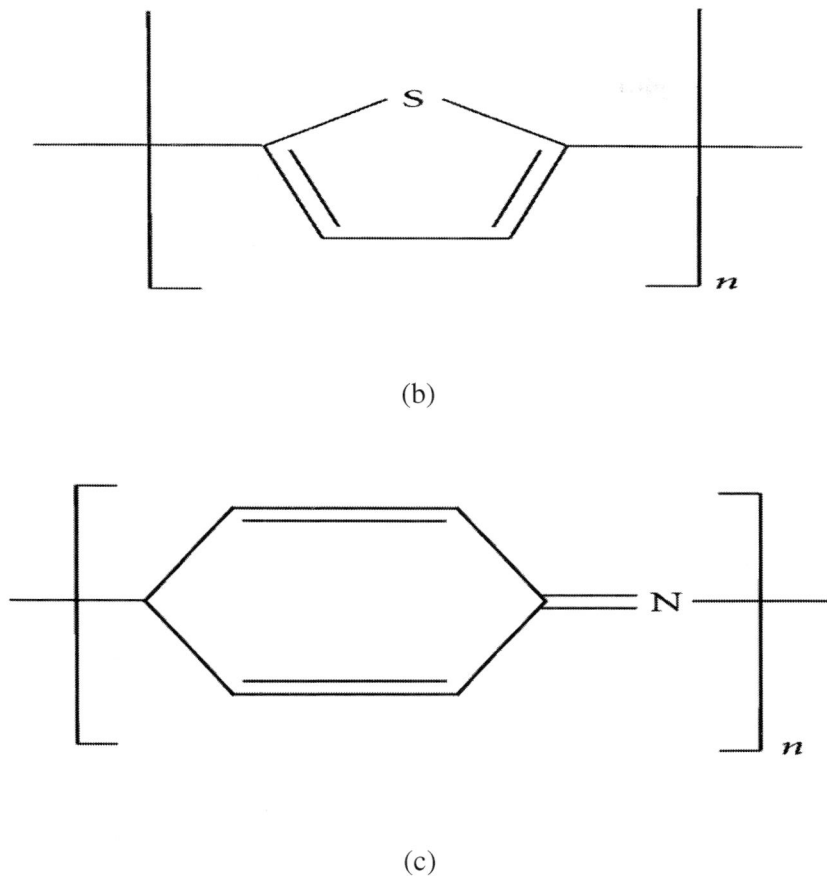

(b)

(c)

Figure 1: Typical conducting polymers: (a) polypyrrole (PPy), (b) polythiophen (PThio), and (c) polyaniline (PAni).

For application of the conducting polymer to corrosion protection, DeBerry was firstly reported in 1985, who presented that the stainless steel covered by PAni was kept in the passive state for relatively long period in sulfuric acid solution [2]. Wessling then pointed out that the conducting polymer coating of polyaniline and polypyrrole possibly possessed self-healing properties, in which the passive oxide between the substrate metal and the conducting polymer could be spontaneously reformed at a flawed site by oxidative capability of the conducting polymer [3].

When anodic potentials are applied to electrodes covered by the conducting polymers shown in Figure 1 after the polymerization, the oxidative property is provided in addition to the conductivity. The ability of the conducting polymer to oxidize the substrate steels allows potential of steels to be shifted to the passive state, in which the steels are protected by the passive oxide formed beneath the conducting polymer. The application of the conducting polymer coating to the corrosion protection of steels was reviewed by Tallman et al. [4]. In this paper, the application of a bilayered conducting PPy to the protection of the steels is reviewed.

CONDUCTING POLYMER

Oxidative polymerization and the doping of anions into the polymer to provide the electronic conductivity have been reviewed by many authors, and here we briefly describe the process of PPy. When the electrode is anodically polarized in an electrolyte solution containing monomer of pyrrole (Py), the black polymer film can be formed on the electrode. The polymerization procedure is done without any difficulty, except for careful treatment of the electrolyte in which oxidation of the Py monomer by air should be avoided. The electrolyte should be thus deoxygenated by inert gas bubbling.

Figure 2 illustrates a model of the process for anodic polymerization of PPy proposed by Genies et al. [5]. Py monomer dissolved in the electrolyte donates an electron into the electrode, resulting in formation of a radical-cation pair (step (1)). The radicals in Py are reacted with each other and two protons are removed from the reacted Py pair (step (2)), forming a dimer of Py (step (3)). After the formation of the radical-cation pair and the reaction between the radicals are repeated, the black PPy film is formed on the electrode (step (4)).

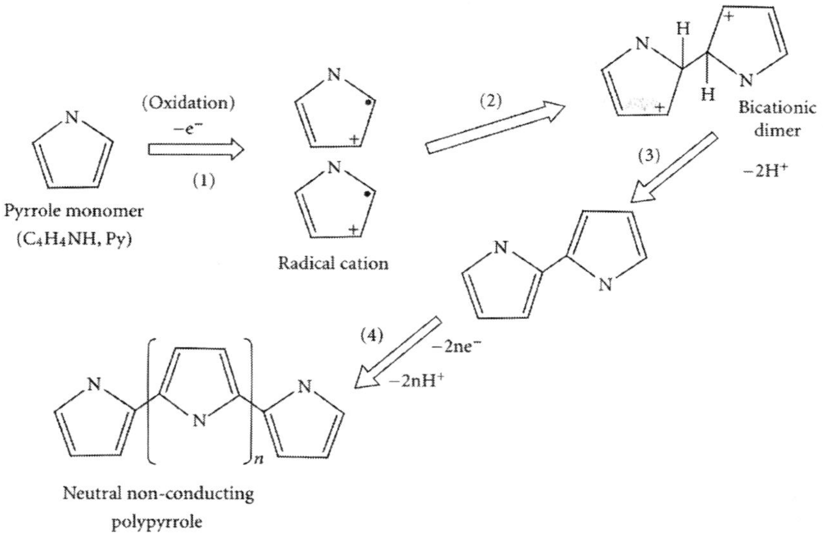

Figure 2: Electropolymerization process of PPy.

The neutral PPy thus formed with a conjugated chain does not possess any conductivity. To add the conductivity into the neutral PPy, further oxidation is required as shown in Figure 3. When the anodic potential is applied to the electrode, an electron is removed from π electrons in the conjugated bond, yielding a pair of a radical and a positive charge (or cation) in the PPy backbone. This situation is called radical-cation state or polaron state. When the two radicals in the PPy are combined, the sites of single and double bond are replaced with each other and two cations remain in the PPy, the situation of which is called bication state or bi-polaron state. The cation thus formed in the PPy can move through π electron clouds, yielding electronic conductivity in the PPy backbone.

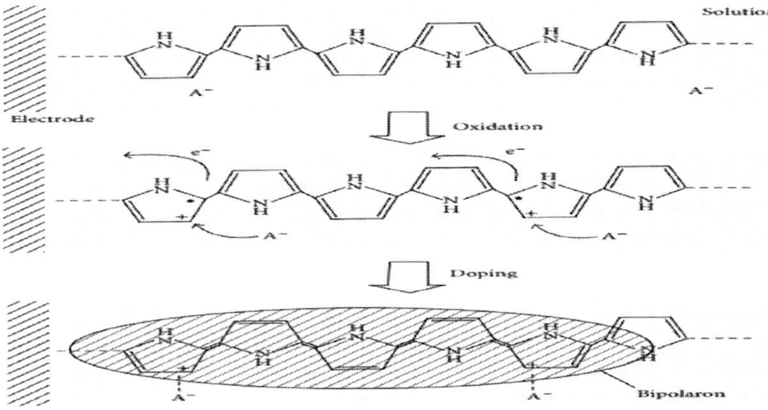

Figure 3: Electrochemical oxidation of neutral nonconducting PPy. During the oxidation, electron transfer from PPy to substrate steel and doping of anions from electrolyte solution to PPy simultaneously occur.

With the removal of electrons from the PPy backbone, insertion of anions from the environmental electrolyte solution occurs to maintain neutrality of the PPy layer; that is, when the neutral state of PPy changes to the oxidative state, removal of electrons and doping of anions simultaneously take place. It is assumed that one positive charge (or cation) can be inserted in three or four Py units at maximum. When more positive charge is added, the PPy changes to overoxidation state and loses the conductivity.

CORROSION PROTECTION OF STEELS BY CONDUCTING POLYMER OF PPY

Mechanism of Corrosion Protection

For the corrosion protection, two mechanisms have been proposed; one is the physical barrier effect, and the other is anodic protection. On the barrier effect, the polymer coating works as a barrier against the penetration of oxidants and aggressive anions, protecting the substrate metals. This effect is similar to paint coating which inhibits the substances from penetrating to the substrate steel. On the anodic protection, the conducting polymer with the strongly oxidative property works as an oxidant to the substrate steel, potential of which is shifted to that in the passive state. In solution at neutral pH, the corrosion potential (or open circuit potential in corrosion) of bare steel is located in the active potential region and the corrosion rate of the steel is usually relatively high. Owing to the coating of conducting polymer, the maximum current in the active-passive transition region was limited by the barrier effect, and then the potential can be easily shifted to the higher potential in the passive state by the strongly oxidative property of the conducting polymer (Figure 4). In the passive state, the corrosion rate of steel becomes much lower. It is assumed that both the barrier effect and the oxidative property induce the anodic protection. Finally, the potential of the substrate steel may be in agreement with a redox potential of the PPy layer in the following reaction, and thus, depends on the degree of oxidation state of the PPy layer.

$$PPy^{n+} \cdot \left(\frac{n}{x}\right) A^{x-} + me$$

$$\rightleftharpoons PPy^{(n-m)+} \cdot \left(\frac{n-m}{x}\right) A^{x-} + \left(\frac{m}{x}\right) A^{x-} \text{aq.}$$

$$(1)$$

The conductivity of the PPy layer affects the oxidative power which brings about the passive state. If the coating layer has little conductivity, the role of the coating as the oxidant is limited in the neighbourhood of the passive oxide. If the layer has enough high conductivity; however, the oxidant power of the whole layer is available and the power increases with the increase of the layer thickness.

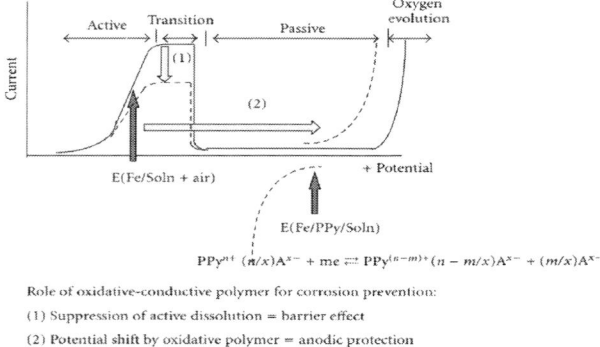

$$PPy^{n+} (n/x)A^{x-} + me \rightleftharpoons PPy^{(n-m)+}(n - m/x)A^{x-} + (m/x)A^{x-}$$

Role of oxidative-conductive polymer for corrosion prevention:
(1) Suppression of active dissolution = barrier effect
(2) Potential shift by oxidative polymer = anodic protection

Figure 4: Potential-current relation of steels covered by oxidative conducting PPy. A barrier effect of PPy suppresses active dissolution of the steel and an oxidative property of PPy shifts the potential into passive state.

The oxidation degree and the conductivity are assumed to decline with longer exposure to environment. If oxidants in the environment reoxidize the degraded PPy layer, the oxidation degree and conductivity can be recovered. When the oxidant in the environment, typically oxygen gas in air, can recover the PPy layer, the duration to maintain the oxidative power of the PPy layer can be prolonged and the passive state of the steel underneath the PPy layer can be kept for a longer period. The recovery process is illustrated in Figure 5.

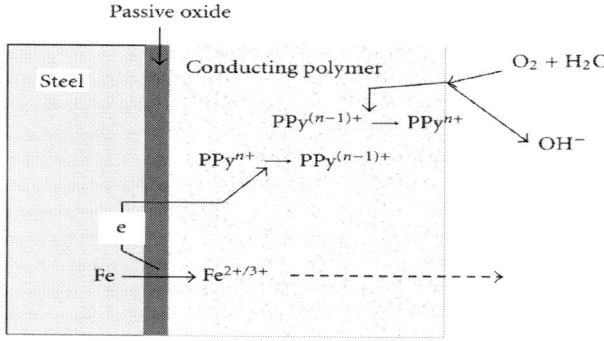

Figure 5: Degradation of oxidative property of PPy and recovery, which was done from reduction of oxygen on the PPy surface.

Ion Exchange in the Conducting Polymer and Its Effect on Corrosion Protection

In the anodic protection, the largest problem is breakdown of passive oxide due to the attack of aggressive anions such as chloride and bromide ions in solution and the breakdown is followed by a large damage of localized corrosion of pitting and crevice corrosion. As contrasted with the cathodic protection, there is a large risk of the localized corrosion connected with the anodic protection. When we control the doping ions in the PPy layer, we possibly prevent penetration of the aggressive anions into the PPy layer. When the steels covered with the conducting PPy are immersed in the sodium chloride solution, the anions doped in the PPy can be exchanged with the chloride anions in the aqueous solution. The chloride anions penetrate the PPy to the substrate steels, and then induce the breakdown of the surface passive oxide film, followed by the pitting corrosion.

The mobility of the dopant anions in the PPy is affected by their mass and volume. When we adopted organic acid ions as the dopant ions in the PPy, they possessed enough large mass and volume to be immobile in the PPy. In general, organic acid anions with large mass are assumed to have small mobility and diffusion in the PPy layer. Accompanied with the oxidation and reduction of the PPy, small anions are doped into and dedoped out of the PPy, respectively, to maintain the neutrality, as described in reaction (1) and shown in Figure 6(a). When the mobility and diffusion of the doped anions are restricted to small value, reversely, the cations are dedoped out of and doped into the PPy during the oxidation and reduction, respectively. The dedoping process of cations in the PPy during the oxidation and the doping during the reduction are described in the following reaction (2):

$$PPy^{n+} \left(\frac{n}{x} \right) B^{x-} + \left(\frac{m}{y} \right) M_{aq}^{y+} + me$$

$$\rightleftarrows PPy^{(n-m)+} \left(\frac{n}{x} \right) B^{x-} \left(\frac{m}{y} \right) M^{y+}.$$

$$(2)$$

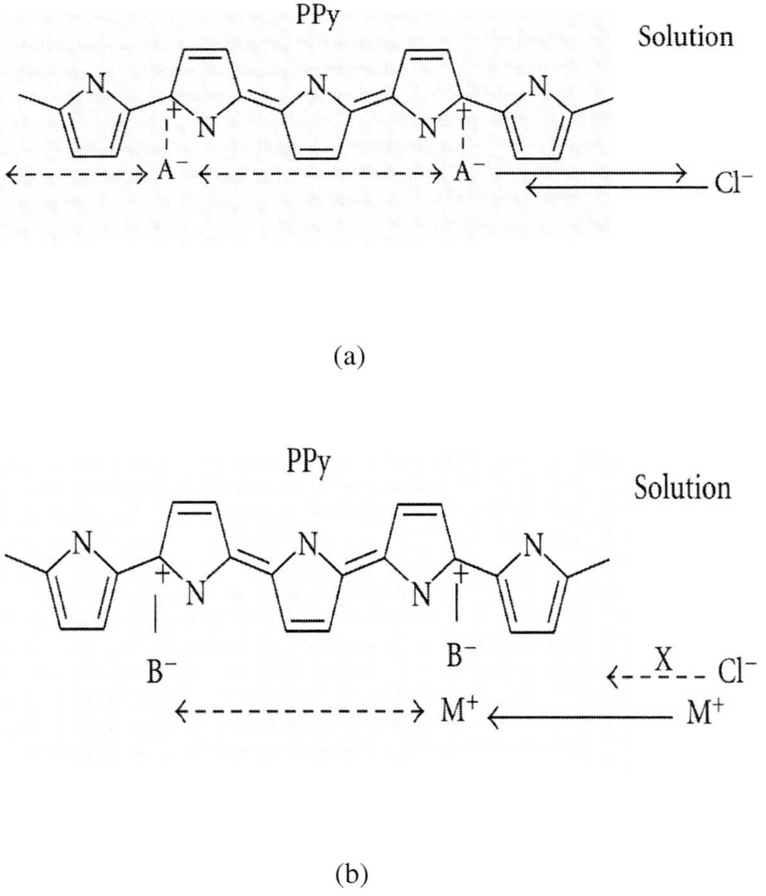

(a)

(b)

Figure 6: Ionic perm-selectivity of PPy film. (a) PPy film with anionic perm-selectivity, in which small-sized anions are doped in PPy and (b) PPy film with cationic perm-selectivity, in which large-sized anions are doped in PPy film.

When one considers the conducting PPy as a charged membrane, the immobile anions with large mass are assumed to have fixed sites with negative charge in the PPy. In the channel between the negatively charged sites, the cations can be mobile and the movement of the anions is greatly inhibited; that is, the membrane exhibits cationic perm-selectivity. As illustrated in Figure 6(b), under the situation where the dopant anions are large enough, the anions in the solution are excluded from the PPy and the substrate steel is protected against the pitting corrosion by chloride attack.

Design for Corrosion Protection by the PPY

The anodic protection greatly depends on the passivity and passive oxide on the steel. For the protection, the passivity and passive oxide must be kept stable. Further, the prevention of penetration of aggressive anions play an important role in the protection.

Deslouis et al. anodically prepared a PPy film on steel from an oxalate solution containing Py monomer and reported that the PPy layer protected the steel in sodium chloride solution for a long period [6–8]. They assumed that the ferric oxalate layer, which was formed underneath the PPy film by the polymerization, worked as a passivation film against corrosion. They also presented that the overcoat layer of PPy doped with dodecylsulfate, $C_{12}H_{25}OSO_3^-$ (DoS), anions was effective to the corrosion protection and that a bilayer coating of PPy-oxalate/PPy-DoS could maintain the passivation state for longer than 500 h, in which no corrosion products were observed.

DoS ion is a surfactant and forms micelle in aqueous solution at concentrations higher than critical concentration. Py monomers, which are probably incorporated in the micelle of DoS in aqueous solution, start to be polymerized when the micelles are collapsed on the electrode to which anodic potential is applied. DoS ions have relatively large masses and work as an immobile dopant in the PPy. The PPy doped with DoS thus is considered as a membrane with negatively charged fixed sites and thus, with cationic perm-selectivity. The outer layer of PPy-DoS can, therefore, exclude the insertion of aggressive anions such as chloride ions.

In Figure 7 the mass change is plotted with anodic oxidation and cathodic reduction of a gold electrode covered with the PPy layers [9]. The mass change was measured by electrochemical quartz crystal microbalance (EQCM) with gold coating. The gold coating was covered by PPy doped with phosphomolybdate ions, $PMo_{12}O_{40}^{3-}$ (PMo) and a bilayered PPy of PPy-PMo/PPy-DoS. The mass change of the PPy-PMo film in Figure 7(a) indicates the uptake of mass during the oxidation and inversely, the removal during the reduction. The behaviour of the mass change during the oxidation reflects the removal of electrons from the PPy and simultaneous insertion of anions from the electrolyte to the PPy and viceversa during the reduction. When one introduces the outer layer of PPy-DoS, the mass change is inversely different from the above result, as shown in Figure 7(b). During the oxidation the mass increases and during the reduction it decreases. In PPy-DoS layer, in which negatively charged ions are fixed, the cations are mobile; during the oxidation the simultaneous removal of both electrons and cations from PPy and during the reduction viceversa. It can be understood that the PPy doped with DoS functions as a cationic perm-selective membrane.

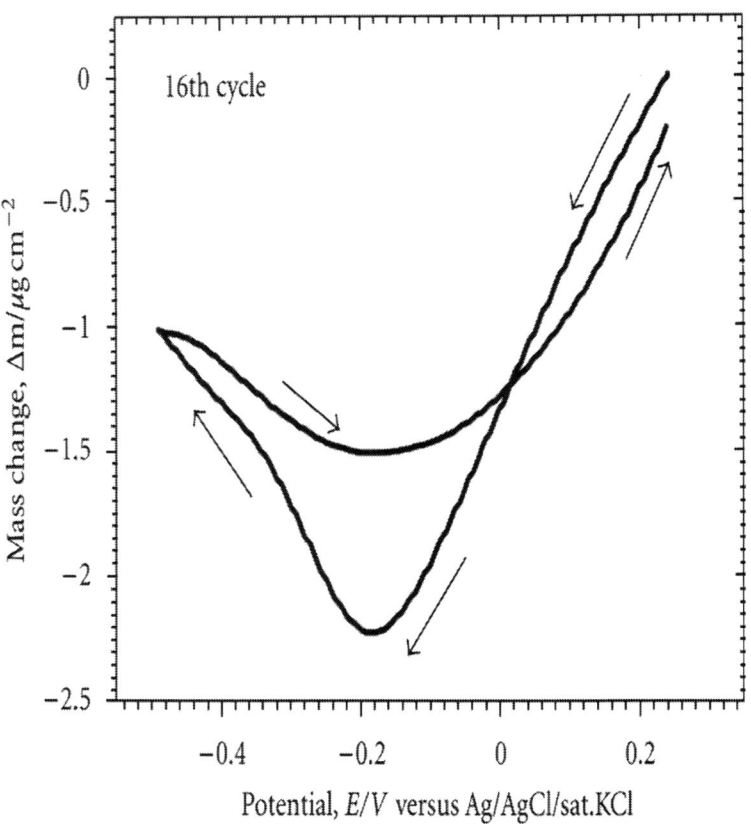

(a)

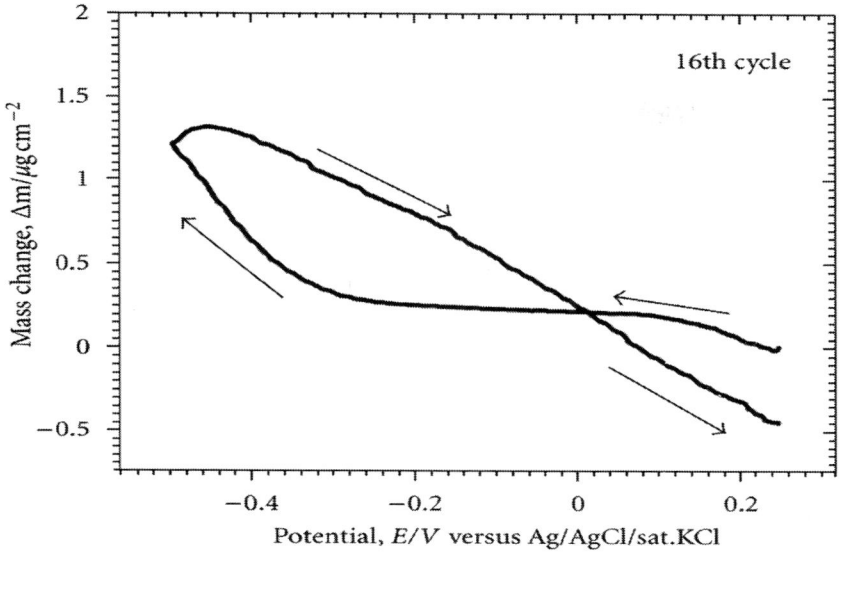

(b)

Figure 7: Mass change of the PPy-PMo layer and bilayer of PPy-PMo/PPy-DoS during the potentiodynamic reduction and oxidation in 3.5% NaCl solution. The data was a result of the 16th redox cycle.

Kowalski et al. designed the corrosion protection PPy layer of steels as following [9–14]. For the inner layer, the PPy was doped with PMo. PMo works as a passivator which stabilizes the passive state of steels and facilitates the formation of passive oxide. For the outer layer, the PPy doped with DoS was prepared. The outer layer can inhibit the anions from penetrating in the PPy layer. The results by Kowalski et al. are shown in Figure 8, [13] where the open circuit potential of the steel covered with the bilayered PPy is plotted during the immersion in 3.5% sodium chloride solution. The steel covered with the bilayered PPy, about 5 μm thick consisting of PPy-PMo/PPy-DoS exhibited the passivation for 190 h in which no corrosion products were observed. If the steel was covered with a single PPy-DoS layer of the same thickness, the passivation is kept for 10 h. It is assumed that PMo ion doped in the inner PPy stabilizes the passive oxide and helps the maintenance of the passive state of the substrate steel.

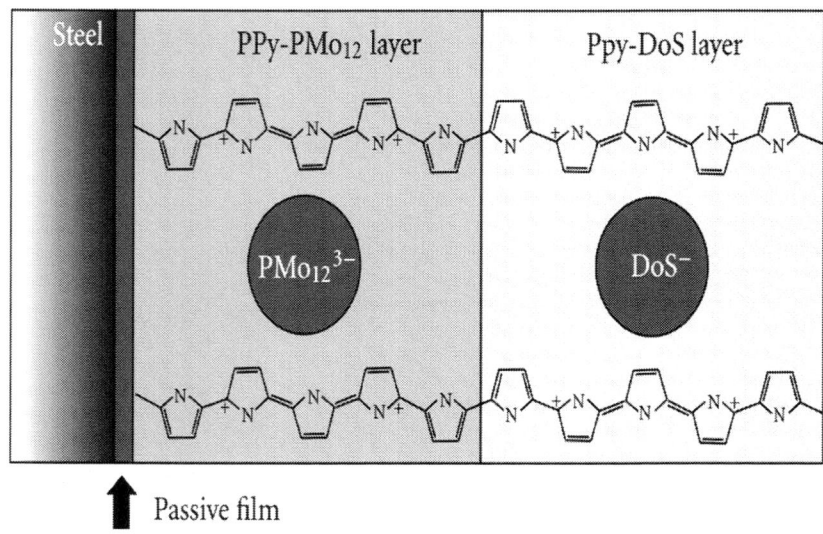

(a)

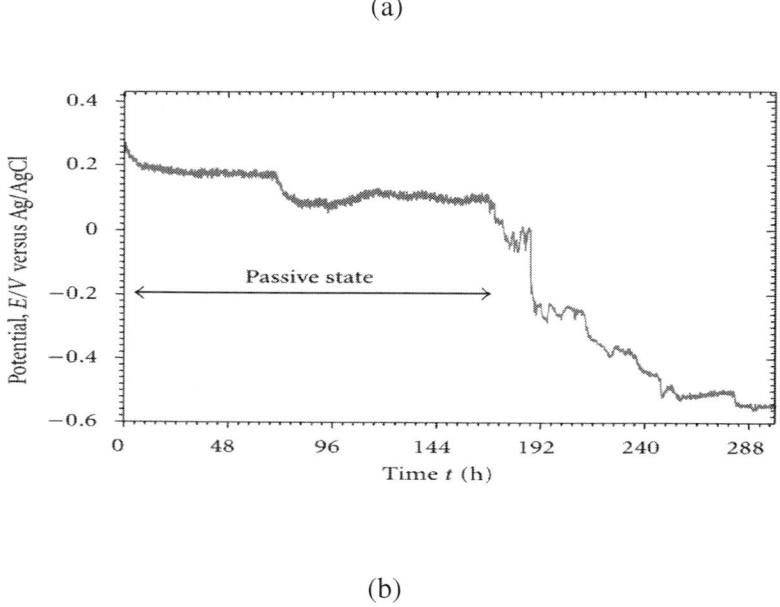

(b)

Figure 8: Model of bilayered PPy film and transient of open circuit potential of steel covered by the bilayered PPy in 3.5% NaCl solution.

The design which combines the inner layer stabilizing the passive oxide with the outer later inhibiting anions from penetrating through PPy to the steel may be suitable to the corrosion protection of steel.

Self-Healing Property

In the corrosion protection, the coating must tolerate small defects to be considered as a suitable replacement for chromate-based coatings. We expect for the conducting polymer coating a self-healing property in which the passive oxide is spontaneously repaired after it develops small defects. On the chromate coating, the chromate ions dissolved from the coating oxidize the steel surface at the damaged sites to reform the passive oxide

$$2Fe + 2CrO_4{}^{2-} + 4H^+ \longrightarrow Fe_2O_3 + Cr_2O_3 + 2H_2O.$$

(3)

A self-healing model proposed by Kowalski et al. is shown in Figure 9 for the bilayered PPy of PPy-PMo/PPy-DoS [9]. After the coating and passive oxide were locally flawed, PMo in the PPy layer is hydrolyzed and decomposes to molybdate and phosphate ions, and then both ions reach the flawed sites. The molybdate ions react with ferric ions on the flawed site to produce the ferric molybdate film. The salt film may be gradually changed to the passive oxide on the damaged site

$$Fe \longrightarrow Fe^{3+} + 3e.$$

(4)

$$2Fe^{3+} + 3MoO_4{}^{2-} \longrightarrow Fe_2(MoO_4)_3.$$

(5)

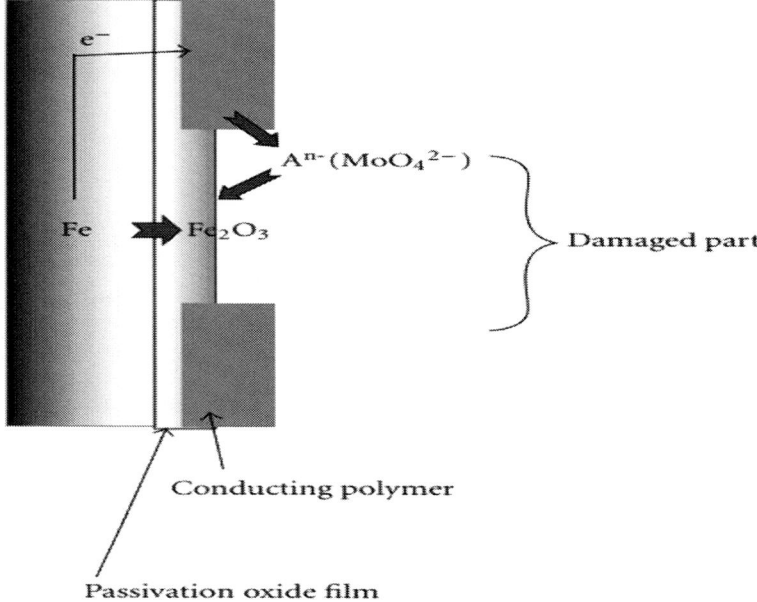

Figure 9: Schematic model of self-healing property of PPy-PMo$_{12}$/PPy-DoS bilayered PPy. Molybdate anions, dissolving from PPy film, reform a passive oxide at the damaged part.

Figure 10 shows the results reported by Kowalski et al. in which a small flaw was inserted by cutting knife in 2 h during the immersion in 3.5% sodium chloride solution [9]. After the PPy layer received the small flaw, the open circuit potential temporarily fell down. When the corrosion continues at the defect site, the potential will decrease to that of bare steel. The potential, however, rose up and recovered in the passive potential region. After that, the potential maintained the high potential in the passive region. When the flawed local site was measured by Raman scattering spectroscopy under this situation, the molybdate salt was detected [9]. It was found that a salt layer of ferric molybdate was reformed on the site.

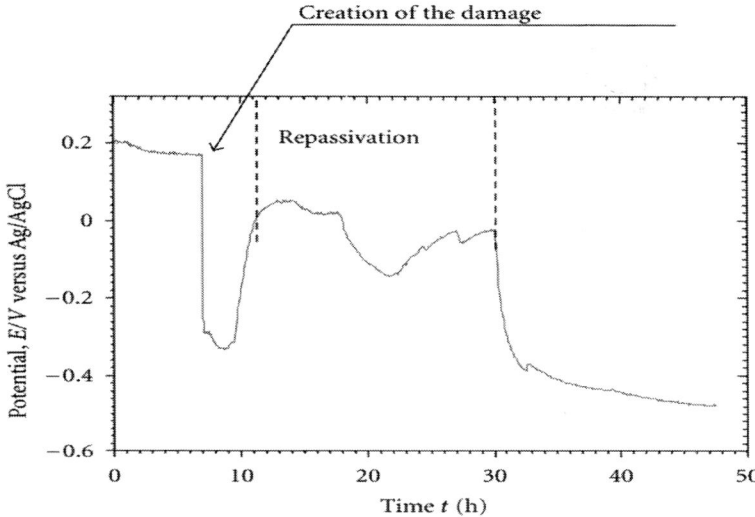

Figure 10: Potential change of steel covered by bilayered PPy film during the immersion in 3.5% NaCl solution. The damage was inserted on the PPy layer by a small knife in 7 h.

SUMMARY

Many papers on the corrosion protection by conducting polymer have been published since 10 years. In those, the attention was paid to how to form homogeneous and adherent layers of conducting polymer on steels and other metals. For the corrosion protection, we must consider the design of the conducting polymer. Since the corrosion protection by the conducting polymer is based on the anodic protection mechanism, we must consider how to stabilize the passive oxide underneath the polymer layer and how to inhibit the aggressive anions from penetrating the polymer layer.

Two mechanisms have been considered for the corrosion protection; one is physical barrier model and the other anodic protection model. We assume that the barrier effect suppresses the active dissolution of steel, facilitating the potential to be shifted in the passive region. The oxidative capability of the conducting polymer helps the potential shift and long maintenance of the passive state of the steel.

Our bilayered model, designed for the corrosion protection, includes two important factors: one is stabilization of the passive film on the steel by action of dopant ions in the inner PPy layer and the other is control of ionic perm-selectivity by organic acid ions doped in the outer PPy layer.

REFERENCES

1.   C. K. Chiang, C. R. Fincher, Y. W. Park et al., "Electrical conductivity in doped polyacetylene," Physical Review Letters, vol. 39, no. 17, pp. 1098–1101, 1977.

2.   D. W. DeBerry, "Modification of the electrochemical and corrosion behavior of stainless steels with an electroactive coating," Journal of the Electrochemical Society, vol. 132, no. 5, pp. 1022–1026, 1985.

3.   B. Wessling, "Passivation of metals by coating with polyaniline: corrosion potential shift and morphological changes," Advanced Materials, vol. 6, no. 3, pp. 226–228, 1994.

4.   D. E. Tallman, G. Spinks, A. Dominis, and G. G. Wallace, "Electroactive conducting polymers for corrosion control: Part 1. General introduction and a review of non-ferrous metals," Journal of Solid State Electrochemistry, vol. 6, no. 2, pp. 73–84, 2002.

5.   E. M. Genies, G. Bidan, and A. F. Diaz, "Spectroelectrochemical study of polypyrrole films," Journal of Electroanalytical Chemistry, vol. 149, no. 1-2, pp. 101–113, 1983.

6.   H. Nguyen Thi Le, B. Garcia, C. Deslouis, and Q. Le Xuan, "Corrosion protection and conducting polymers: polypyrrole films on iron," Electrochimica Acta, vol. 46, no. 26-27, pp. 4259–4272, 2001.

7.   N. T. L. Hien, B. Garcia, A. Pailleret, and C. Deslouis, "Role of doping ions in the corrosion protection of iron by polypyrrole films," Electrochimica Acta, vol. 50, no. 7-8, pp. 1747–1755, 2005.

8.   T. Van Schaftinghen, C. Deslouis, A. Hubin, and H. Terryn, "Influence of the surface pre-treatment prior to the film synthesis, on the corrosion protection of iron with polypyrrole films," Electrochimica Acta, vol. 51, no. 8-9, pp. 1695–1703, 2006.

9. D. Kowalski, M. Ueda, and T. Ohtsuka, "Self-healing ion-permselective conducting polymer coating,"Journal of Materials Chemistry, vol. 20, no. 36, pp. 7630–7633, 2010.

10. T. Ohtsuka, M. Iida, and M. Ueda, "Polypyrrole coating doped by molybdo-phosphate anions for corrosion prevention of carbon steels," Journal of Solid State Electrochemistry, vol. 10, no. 9, pp. 714–720, 2006.

11. D. Kowalski, M. Ueda, and T. Ohtsuka, "Corrosion protection of steel by bi-layered polypyrrole doped with molybdophosphate and naphthalenedisulfonate anions," Corrosion Science, vol. 49, no. 3, pp. 1635–1644, 2007.

12. D. Kowalski, M. Ueda, and T. Ohtsuka, "The effect of counter anions on corrosion resistance of steel covered by bi-layered polypyrrole film," Corrosion Science, vol. 49, no. 8, pp. 3442–3452, 2007.

13. D. Kowalski, M. Ueda, and T. Ohtsuka, "The effect of ultrasonic irradiation during electropolymerization of polypyrrole on corrosion prevention of the coated steel," Corrosion Science, vol. 50, no. 1, pp. 286–291, 2008.

14. D. Kowalski, M. Ueda, and T. Ohtsuka, "Self-healing ability of conductive polypyrrole coating with artificial defect," ECS Transactions, vol. 16, no. 52, pp. 177–182, 2008.

Citations

CHAPTER 1

Andréa Santos Liu and Maria Auxiliadora Silva Oliveira, Corrosion Control of Aluminum Surfaces by Polypyrrole Films: Influence of Electrolyte, doi.org/10.1590/S1516-14392007000200018.

CHAPTER 2

Matthew D. Pritzl, Habib Tabatabai, and Al Ghorbanpoor, "Laboratory Assessment of Select Methods of Corrosion Control and Repair in Reinforced Concrete Bridges," International Journal of Corrosion, vol. 2014, Article ID 175094, 11 pages, 2014, doi:10.1155/2014/175094.

CHAPTER 3

Neha Patni, Shruti Agarwal, and Pallav Shah, "Greener Approach towards Corrosion Inhibition,"Chinese Journal of Engineering, vol. 2013, Article ID 784186, 10 pages, 2013. doi:10.1155/2013/784186.

CHAPTER 4

B. E. Amitha Rani and Bharathi Bai J. Basu, "Green Inhibitors for Corrosion Protection of Metals and Alloys: An Overview," International Journal of Corrosion, vol. 2012, Article ID 380217, 15 pages, 2012. doi:10.1155/2012/380217.

CHAPTER 5

El-Sayed M. Sherif, Adel Taha Abbas, D. Gopi, and A. M. El-Shamy, "Corrosion and Corrosion Inhibition of High Strength Low Alloy Steel in 2.0 M Sulfuric Acid Solutions by 3-Amino-1,2,3-triazole as a Corrosion Inhibitor," Journal of Chemistry, vol. 2014, Article ID 538794, 8 pages, 2014. doi:10.1155/2014/538794.

CHAPTER 6

Ambrish Singh, Eno E. Ebenso, and M. A. Quraishi, "Corrosion Inhibition of Carbon Steel in HCl Solution by Some Plant Extracts," International Journal of Corrosion, vol. 2012, Article ID 897430, 20 pages, 2012. doi:10.1155/2012/897430.

CHAPTER 7

I. Y. Suleiman, O. B. Oloche, and S. A. Yaro, "The Development of a Mathematical Model for the Prediction of Corrosion Rate Behaviour for Mild Steel in 0.5 M Sulphuric Acid," ISRN Corrosion, vol. 2013, Article ID 710579, 9 pages, 2013. doi:10.1155/2013/710579.

CHAPTER 8

Benjamin Valdez Salas, Michael Schorr Wiener, Roumen Zlatev Koytchev, et al., "Copper Corrosion by Atmospheric Pollutants in the Electronics Industry," ISRN Corrosion, vol. 2013, Article ID 846405, 7 pages, 2013. doi:10.1155/2013/846405.

CHAPTER 9

Toshiaki Ohtsuka, "Corrosion Protection of Steels by Conducting Polymer Coating," International Journal of Corrosion, vol. 2012, Article ID 915090, 7 pages, 2012. doi:10.1155/2012/915090.

Index